Principles of Computer Graphics

Theory and Practice Using OpenGL and Maya®

Principles of Computer Graphics

Theory and Practice Using OpenGL and Maya®

Shalini Govil-Pai
Sunnyvale, CA, U.S.A.

Shalini Govil-Pai
896 Savory Drive,
Sunnyvale, CA 94087
Email: sgovil@gmail.com

Library of Congress Cataloging-in-Publication Data

Govil-Pai, Shalini
Principles of Computer Graphics: Theory and Practice Using OpenGL and Maya®/ Shalini Govil-Pai
p. cm.
Includes bibliographical references and index.

ISBN 0-387-95504-6 (HC) e-ISBN 0-387-25479-X
ISBN-13: 978-0387-95504-9 e-ISBN-13: 978-0387-25479-1

---

springer.com

Govil-Pai: Principles of Computer Graphics: Theory and Practice Using OpenGL and Maya®

First Indian Reprint 2007
Third Indian Reprint 2008

ISBN 978-81-8128-674-1

This edition is manufactured in India for sale only in India, Pakistan, Bangladesh, Nepal and Sri Lanka and any other country as authorized by the publisher.

This edition is published by Springer (India) Private Limited,
A part of Springer Science+Business Media, Registered Office: 906-907,
Akash Deep Building, Barakhamba Road, New Delhi – 110 001, India.

Printed in India by Raj Press, New Delhi.

Contents

Preface

Computer Graphics: the term has become so widespread now, that we rarely stop to think about what it means. What is Computer Graphics? Simply defined, Computer Graphics (or CG) is the images generated or modified on a computer. These images may be visualizations of real data or imaginary depictions of a fantasy world.

The use of Computer Graphics effects in movies such as *The Incredibles* and games such as *Myst* have dazzled millions of viewers worldwide. The success of such endeavors is prompting more and more people to use the medium of Computer Graphics to entertain, to educate, and to explore.

For doctors, CG provides a noninvasive way to probe the human body and to research and discover new medications. For teachers, CG is an excellent tool to visually depict concepts to their students. For business people, CG has come to signify images of charts and graphs used for analysis of data. But for most of us, CG translates into exciting video games, special effects and entire films-what are often referred to as CG productions. This entertainment aspect of CG is what has made it such a glamorous and sought-after field.

Ten years ago, CG was limited to high-end workstations, available only to an elite few. Now, with the advances in PC processing power and the availability of 3D graphics cards, even high school students can work on their home PC to create professional quality productions

The goal of this book is to expose you to the fundamental principles behind modern computer graphics. We present these principles in a fun and simple manner. We firmly believe that you don't have to be a math whiz or a high tech computer programmer to understand CG. A basic knowledge of trigonometry, algebra, and computer programming is more than sufficient.

As you read this book, you will learn the bits and bytes of how to transform your ideas into stunning visual imagery. We will walk you through the processes that professionals employ to create their productions, Based on the principles that we discuss, you will follow these processes step and step, to design and create your own games and animated movies.

We will introduce you to the OpenGL API—a graphics library that has become the de facto standard on all desktops. We will also introduce you to the workings of Maya, a 3D software package. We will demonstrate the workings of the Maya Personal Learning Edition—a (free) download is required.

Organization of the Book

The book is organized into three sections. Every section has detailed OpenGL code and examples. Appendix B details how to install these examples on your desktop.

Section 1: The Basics

The first section introduces the most basic graphics principles. In Chapter 1, we discuss how the computer represents color and images. We discuss how to describe a two-dimensional (2D) world, and the objects that reside in this world. Moving objects in a 2D world involves 2D transformations. Chapter 2 describes the principles behind transformations and how they are used within the CG world. Chapter 3 discusses how the computer saves images and the algorithms used to manipulate these images. Finally, in Chapter 4, we combine all the knowledge from the previous chapters to create our very own version of an arcade game.

Section 2: It's 3D time

Section 2 will expand your horizon from the 2D world to the 3D world. The 3D world can be described very simply as an extension of the 2D world. In Chapter 5, we will introduce you to 3D modeling. Chapter 6 will discuss rendering: you will have the opportunity to render your models from Chapter 5 to create stunning visual effects. Chapter 7 is an advanced chapter for those interested in more advance concepts of CG. We will introduce the concept of Nurbs as used in modeling surfaces. We will also introduce you to advanced shading concepts such as ray tracing. Chapter 8 focuses on teaching the basics of Maya and the Maya Personal Learning Edition of Maya (Maya PLE). Maya is the most popular software in the CG industry and is extensively used in every aspect of production. Learning the basics of this package will be an invaluable tool for those interested in pursuing this area further.

Section 3: Making Them Move

Section 3 discusses the principles of animation and how to deploy them on the computer. In Chapter 9, we discuss the basic animation techniques. Chapter 10 discusses a mode of animation commonly deployed in games, namely, viewpoint animation. In Chapter 11, you will have the opportunity to combine the working knowledge from the previous chapters to create your own movie using Maya.

Appendices

In Appendix A, you will find detailed instructions on how to install the OpenGL and GLUT libraries. Appendix B describes how to download, install the sample code that is detailed in this book. You will also find details on how to compile and link your code using the OpenGL libraries. Appendix C describes the Maya PLE and how to download it.

OpenGL and Maya

Every concept discussed in the book is followed by examples and exercises using C and the OpenGL API. We also make heavy use of the GLUT library, which is a cross-platform OpenGL utility toolkit. The examples will enable you to visually see and appreciate the theory explained. We do not expect you to know OpenGL, but we do expect basic knowledge in C and C++ and knowledge of compiling and running these programs. Some chapters detail the workings of Maya, a popular 3D software package. Understanding Maya will enable you to appreciate the power of the CG concepts that we learn in the book.

Why are we using OpenGL and GLUT?

OpenGL is now a widely accepted industry standard and is used by many (if not all) professional production houses. It is not a programming language but an API. That is, it provides a library of graphics functions for you to use within your programming environment. It provides all the necessary communication between your software and the graphics hardware on your system.

GLUT is a utility library for cross-platform programming. Although our code has been written for the Windows platform, GLUT makes it easier to compile the example code on other platforms such as Linux or Mac. GLUT also eliminates the need to understand basic Windows programming so that we can focus on graphics issues only.

Why are we using Maya?

Some concepts in the book will be further illustrated with the help of industry leading 3D software Maya. Academy-Award winning Maya 3D animation and effects software has been inspired by the film and video artists, computer game developers, and design professionals who use it daily to create engaging digital imagery, animation, and visual effects. Maya is used in almost every production house now, so learning the basics of it will prove to be extremely useful for any CG enthusiast. In addition, the good folks at Alias now let you download a free version of Maya (Maya PLE) to use for learning purposes.

The system requirements for running the examples in this book, as well as for running Maya PLE are as follows:

Software Requirements

- Windows 2000 or higher
- C/C++ compiler such as Microsoft Visual Studio on Windows or GCC (Gnu Compiler Collection) on Unix

Hardware Requirements

- Intel Pentium II or higher/AMD Athlon processor
- 512 MB RAM
- Hardware-accelerated graphics card (comes standard on most systems)

In addition, we expect some kind of Internet connectivity so that you can download required software.

Intended Audience

This book is aimed at undergraduate students who wish to gain an overview of Computer Graphics. The book can be used as a text or as a course supplement for a basic Computer Graphics course.

The book can also serve as an introductory book for hobbyists who would like to know more about the exciting field of Computer Graphics, and to help them decide if they would like to pursue a career in it.

Acknowledgments

The support needed to write and produce a book like this is immense. I would like to acknowledge several people who have helped turn this idea into a reality, and supported me through the making of it:

First, my husband, Rajesh Pai, who supported me through thick and thin. You have been simply awesome and I couldn't have done it without your constant encouragement.

A big thanks to my parents, Anuradha and Girjesh Govil, who taught me to believe in myself, and constantly egged me on to publish the book.

Thanks to Carmela Bourassa of Alias software, who helped provide everything I needed to make Maya come alive.

A very special thanks to my editor, Wayne Wheeler, who bore with me through the making of this book and to the entire Springer staff who helped to produce this book in its final form.

I would like to dedicate this book to my kids, Sonal and Ronak Pai, who constantly remind me that there is more to life than CG.

CG technology is emerging and changing every day. For example, these days, *sub-division surfaces*, *radiosity*, and *vertex shaders* are in vogue. We cannot hope to cover every technology in this book. The aim of the book is to empower you with the basics of CG-providing the stepping-stone to pick up on any CG concept that comes your way.

A key tenet of this book is that computer graphics is fun. Learning about it should be fun too. In the past 30 years, CG has become pervasive in every aspect of our lives. The time to get acquainted with it is now—so read on!

Section 1

The Basics

Imagine how the world would be if computers had no way of drawing pictures on the screen. The entire field of Computer Graphics—flight simulators, CAD systems, video games, 3D movies—would be unavailable. Computers would be pretty much what they were in the 1960s - just processing machines with monitors displaying text in their ghostly green displays.

Today, computers do draw pictures. It's important to understand how computers actually store and draw graphic images. The process is very different from the way people do it. First, there's the problem of getting the image on the screen. A computer screen contains thousands of little dots of light called pixels. To display a picture, the computer must be able to control the color of each pixel. Second, the computer needs to know how to organize the pixels into meaningful shapes and images. If we want to draw a line or circle on the screen, how do we get the computer to do this?

The answers to these questions form the basis for this section. You will learn how numbers written in the frame buffer control the colors of the pixels on the screen. We will expose you to the concept of two-dimensional coordinate systems and how 2D shapes and objects can be drawn and transformed in this 2D world. You will learn the popular algorithms used to draw basic shapes such as lines and circles on the computer.

These days, three-dimensional graphics is in vogue. As a reader, you too must be eager to get on to creating gee-whiz effects using these same principles. It is important, however, to realize that all 3D graphics principles are actually extensions of their 2D counterparts. Understanding concepts in a 2D world is much easier and is the best place to begin your learning. Once you have mastered 2D concepts, you will be able to move on to the 3D world easily. At every step, you will also have the opportunity to implement the theory discussed by using OpenGL.

At the end of the section, we shall put together everything we have learned to develop a computer game seen in many video arcades today.

Chapter 1

From Pixels to Shapes

The fundamental building block of all computer images is the picture element, or the *pixel*. A pixel is a dot of light on the computer screen that can be set to different colors. An image displayed on the computer, no matter how complex, is always composed of rows and columns of these pixels, each set to the appropriate color and intensity. The trick is to get the right colors in the right places.

Since computer graphics is all about creating images, it is fitting that we begin our journey into the computer graphics arena by first understanding the pixel. In this chapter, we will see how the computer represents and sets pixel colors and how this information is finally displayed onto the computer screen. Armed with this knowledge, we will explore the core graphics algorithms used to draw basic shapes such as lines and circles.

In this chapter, you will learn the following concepts:

- What pixels are
- How the computer represents color
- How the computer displays images
- The core algorithm used to draw lines and circles
- How to use OpenGL to draw shapes and objects

1.1 Computer Display Systems

The computer display, or the monitor, is the most important device on the computer. It provides visual output from the computer to the user. In the Computer Graphics context, the display is everything. Most current personal computers and workstations use *Cathode Ray Tube* (CRT) technology for their displays.

As shown in Fig.1.1, a CRT consists of

- An electron gun that emits a beam of electrons (cathode rays)
- A deflection and focusing system that directs a focused beam of electrons towards specified positions on a phosphorus-coated screen
- A phosphor-coated screen that emits a small spot of light proportional to the intensity of the beam that hits it

The light emitted from the screen is what you see on your monitor.

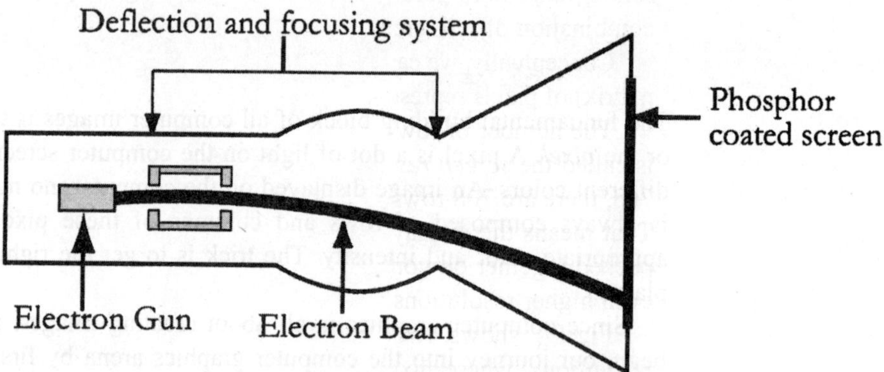

Fig.1.1: A cathode ray tube

The point that can be lit up by the electron beam is called a pixel. The intensity of light emitted at each pixel can be changed by varying the number of electrons hitting the screen. A higher number of electrons hitting the screen will result in a brighter color at the specified pixel. A grayscale monitor has just one phosphor for every pixel. The color of the pixel can be set to black (no electrons hitting the phosphor), to white (a maximum number of electrons hitting the phosphor), or to any gray range in between. A higher number of electrons hitting the phosphor results in a whiter-colored pixel.

A color CRT monitor has three different colored phosphors for each pixel. Each pixel has red, green, and blue-colored phosphors arranged in a triangular group. There are three electron guns, each of which generates an electron beam to excite one of the phosphor dots, as shown in Fig.1.2. Depending on the monitor manufacturer, the pixels themselves may be round dots or small squares, as shown in Fig.1.3.

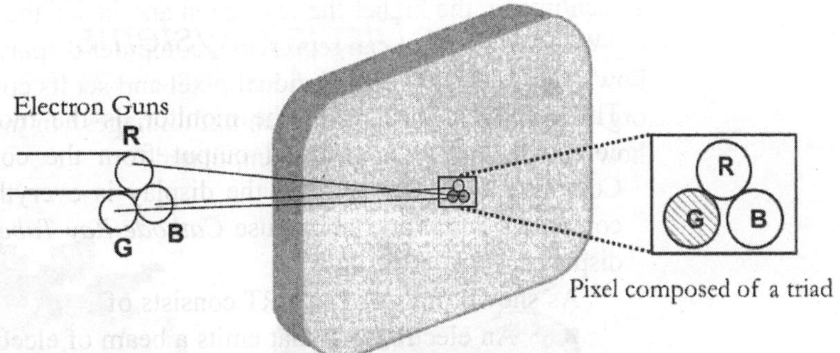

Fig.1.2: Color CRT uses red green and blue triads

Because the dots are close together, the human eye fuses the three red, green, and blue dots of varying brightness into a single dot/square that appears to be the color combination of the three colors. (For those of us who missed art class in school, all colors perceived by humans can be formed by the right brightness combination of red, green, and blue color.)

Conceptually, we can think of the screen as a discrete two-dimensional array (a matrix) of pixels representing the actual layout of the screen, as shown in Fig.1.3.

The number of rows and columns of pixels that can be shown on the screen is called the *screen resolution*. On a display device with a resolution of 1024 x 768, there are 768 rows (scan lines), and in each scan line there are 1024 pixels. That means the display has 768 x 1024=786,432 pixels! That is a lot of pixels packed together on your 14-inch monitor. Higher-end workstations can achieve even higher resolutions.

Fig.1.4 shows two images displayed in different resolutions. At lower resolutions, where pixels are big and not so closely packed, you can start to notice the "pixelated" quality of the image as in the image shown on the right. At higher resolutions, where pixels are packed close together, your eye perceives a smooth image. This is why the resolution of the display (and correspondingly that of the image) is such a big deal.

You may have heard the term *dpi*, which stands for *dots per inch*. The word *dot* is really referring to a pixel. The higher the number of dots per inch of the

Fig.1.3: Computer display: rows and columns of pixels

screen/image, the higher the resolution and hence the crisper the image.

We have seen we can represent a computer display as a matrix of pixels. But how can we identify an individual pixel and set its color? And how can we then organize the pixels to form meaningful images? In the next section, we explore how pixel colors are set and manipulated.

Fig.1.4: The same image at different reolsutions

1.2 Frame Buffers

The light on the screen generated by the beam of electrons in our CRT fades quickly—in 10 to 60 microseconds. In order to keep a picture on the screen for a while, the picture needs be redrawn before it disappears off the screen. This is called *refreshing* the screen. Most display systems use raster scan technology to perform the refresh process. In this technology, the electron beam is directed discretely across the screen, one row at a time from left to right, starting at the upper left corner of the screen. When the beam reaches the bottommost row, the process is repeated, effectively refreshing the screen.

Raster scan systems use a memory buffer called frame buffer (or refresh buffer) in which the intensities of the pixels are stored. Refreshing the screen is performed using the information stored in the frame buffer. You can think of frame buffer as a two dimensional array. Each element of the array keeps the

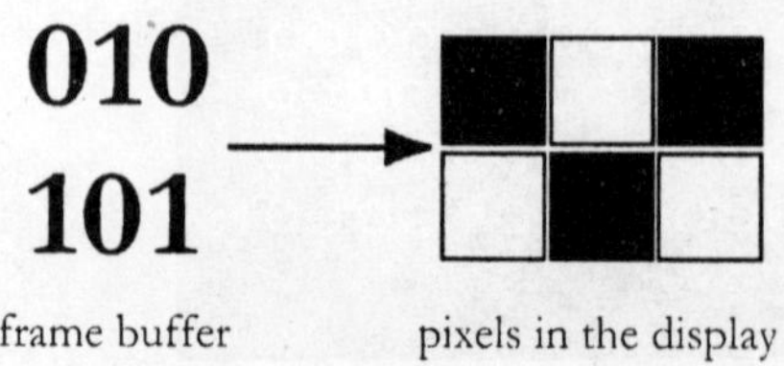

Fig.1.5: Monochrome display: frame buffer for turning pixels on and off

intensity of the pixel on the screen corresponding to that element.

For a monochrome display, the frame buffer has one bit for each pixel. The display controller keeps reading from the frame buffer and turns on the electron gun only if the bit in the buffer is as shown in Fig.1.5.

Systems can have multiple buffers. Foreground buffers draw directly into the window specified. Sometimes a background buffer is also used. The background buffer is not displayed on the screen immediately. We shall talk about buffering modes in more detail when we study animation.

How about color?

You may recall from school physics that all colors in the world can be represented by mixing differing amounts of the three primary colors, namely, red, green, and blue. In CG, we represent color as a triplet of the **R**ed, **G**reen, and **B**lue components. The triplet defines the final color and intensity. This is called the *RGB color model.* Color Plate 1 shows an image of Red, Green and Blue circles and the resultant colors when they intersect.

Some people use a minimum of 0 and a maximum of 255 to represent the intensities of the three primaries, and some people use a floating-point number between 0 and 1. In this book (as is the case in OpenGL), we shall use 0 to represent no color and 1.0 to represent the color set to its maximum intensity. Varying the values in the RGB triplet yields a new color. Table 1.1 lists the RGB components of common colors.

On color systems, each pixel element in the frame buffer is represented by an RGB triplet. This triplet controls the intensity of the electron gun for each of the red, green, and blue phosphors, respectively of the actual pixel on the screen. Our eye perceives the final pixel color to be the color combination of the three colors.

Each pixel color can be set independent of the other pixels. The total number of colors that can be displayed on the screen at one time, however, is limited by the number of bits used to represent color. The number of bits used is called the *color resolution* of the monitor.

For lower resolution systems like VGA monitors, the color resolution is

Color	Red	Green	Blue
Yellow	1	1	0
White	1	1	1
Black	0	0	0
Cyan	0	1	1
Magenta	1	0	1

Table 1.1: The RGB components of common colors

usually 8 bits. Eight-bit systems can represent up to 256 colors at any given time. These kinds of systems maintain a color table. Applications use an index (from 1 to 256) into this color table to define the color of the screen pixel. This mode of setting colors is called *color index mode* and is shown in Fig.1.6.

Of course, if we change a color in this table, any application that indexes into this table will have its color changed automatically. Not always a desirable effect!

Most modern systems have a 24-bit color resolution or higher. A 24-bit system (8 bits for the red channel, 8 for the green channel, and 8 for the blue channel) can display 16 million colors at once. Sometimes an additional 8 bits is added, called the *alpha channel*. We shall look into this alpha channel and its uses later when we learn about fog and blending.

With so many colors available at any given time, there is no need for a color table. The colors can be referred to directly by their RGB components. This way

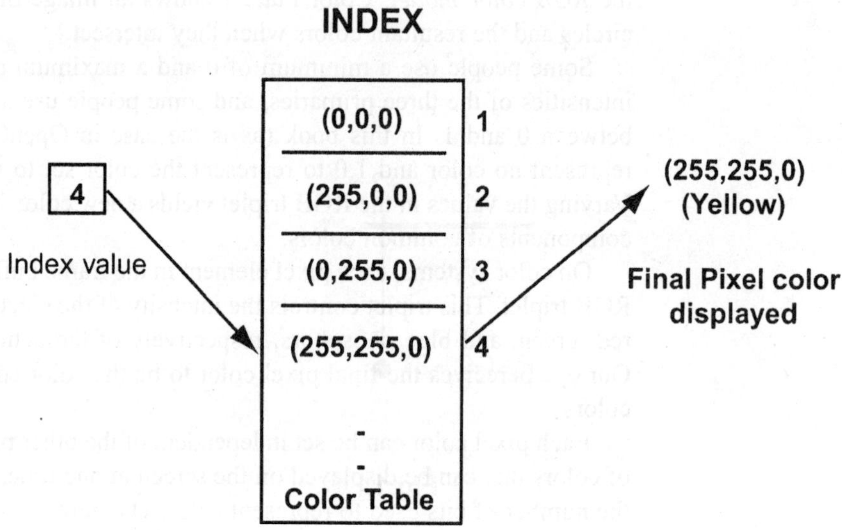

Fig. 1.6:Color index mode

of referring to colors is called *RGB mode*. We shall employ the RGB mode to refer to color throughout the rest of this book.

We have seen how pixel colors are stored and displayed on the screen. But we still need to be able to identify each pixel in order to to set its color. In the next section, we shall see how to identify individual pixel points that we want to paint.

1.3 Coordinate Systems: How to Identify Pixel Points

Coordinates are sets of numbers that describe position—position along a line, a surface of a sphere, etc. The most common coordinate system for plotting both two dimensional and three-dimensional data is the Cartesian coordinate system. Let us see how to use this system to identify point positions in 2D space.

The Cartesian coordinate system is based on a set of two straight lines called the *axes*. The axes are perpendicular to each other and meet at the origin. Each axis is marked with the distances from the origin. Usually an arrow on the axis indicates positive direction. Most commonly, the horizontal axis is called the *x*-axis, and the vertical axis is called the *y*-axis.

Fig.1.7 shows a Cartesian coordinate system with an *x*- and a *y*-axis. To define any point *P* in this system, we draw two lines parallel to the *xy*-axes. The

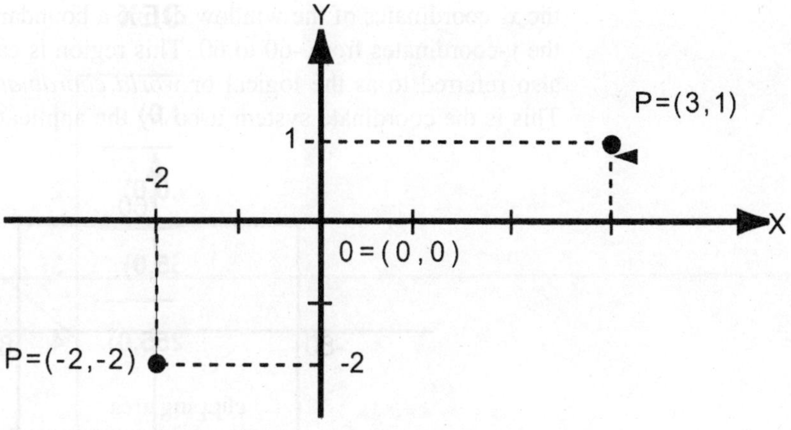

Fig.1.7: Cartesian coordinate system

values of *x* and *y* at the intersections completely define the position of this point. In the Cartesian coordinate system, we label this point as (*x*,*y*). x and y are called the *coordinates* of the point (and could have a negative value). In this system, the origin is represented as (0,0), since it is at 0 distance from itself. A coordinate system can be attached to any space within which points need to be located.

Coming back to the world of computers, recall that our computer display is represented physically in terms of a grid of pixels. This grid of pixels can be defined within its own Cartesian coordinate system. Typically, it is defined with an origin at the upper left corner of the screen. We refer to this Cartesian space as the *physical coordinate system*. Within this system, each pixel can then be uniquely identified by its (*x*,*y*) coordinates, as shown in Fig.1.8.

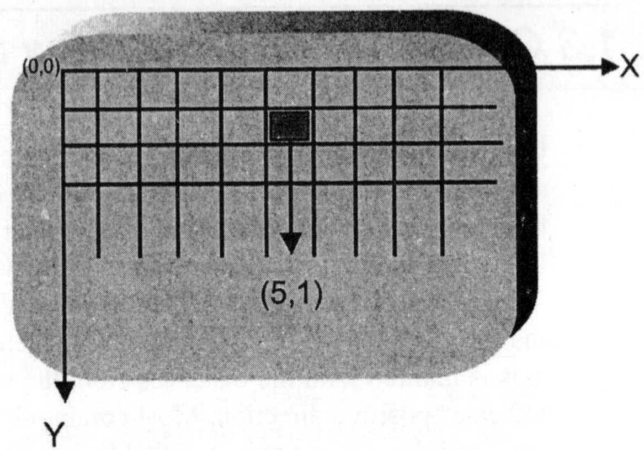

Fig.1.8: Identifying pixels on the screen

Now, consider an application window being shown on this display. We can specify a Cartesian space within which the application resides within. In Fig.1.9, the *x*- coordinates of the window define a boundary ranging from -80 to 80 and the *y*-coordinates from -60 to 60. This region is called the *clipping area*, and is also referred to as the logical or *world coordinate system* for our application. This is the coordinate system used by the application to plot points, draw lines

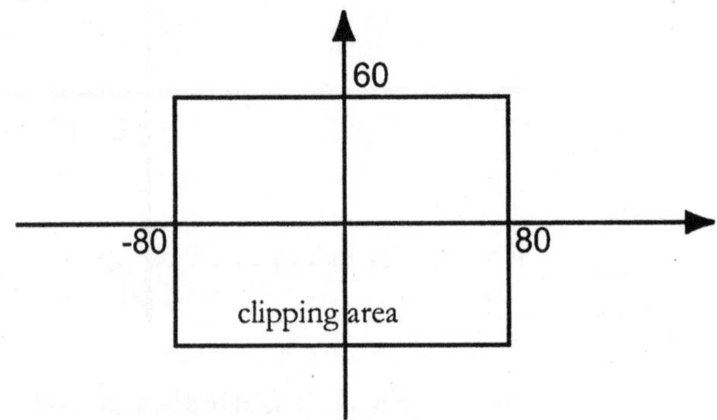

Fig.1.9: Application (clipping) area

and shapes, etc. Objects within the clipping area are drawn, and those outside this area are removed or *clipped* from the scene. The clipping area is mapped onto a physical region in the computer display by mapping the application boundaries to the physical pixel boundaries of the window.

If the clipping area defined matches the (physical) resolution of the window, then each call to draw an (*x*,*y*) point (with integer values) in the world coordinate system will have a one-to-one mapping with a corresponding pixel in the physical coordinate system.

For most applications, the clipping area does not match the physical size of the window. In this case, the graphics package needs to perform a transformation

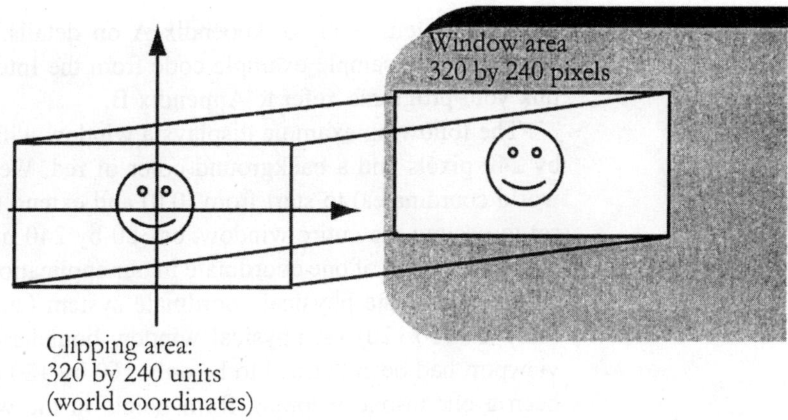

Fig.1.10: Mapping Clipping Area onto the window

from the world coordinate system being used by the application to the physical window coordinates. This transformation is determined by the clipping area, the physical size of the window, and another setting known as the *viewport*. The viewport defines the area of the window that is actually being used by the application.

For now, we will assume that the viewport is defined as the entire window (but this is not always necessary, as shown in Fig.1.11).

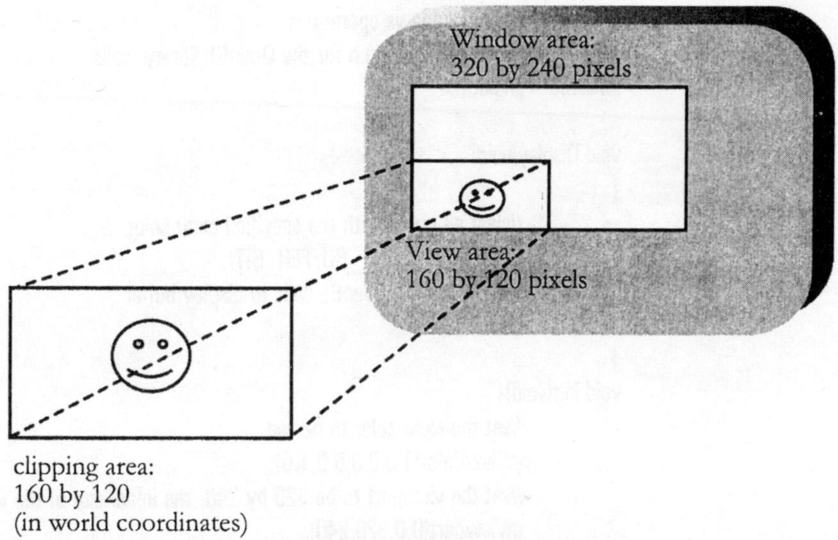

Fig.1.11: Viewport of a Window

Example Time

Let us run our first OpenGL program to get a handle on some of the concepts we have just learned. For information on OpenGL and GLUT and how to install it on your system refer to Appendix A on details. For information on how to download the sample example code from the Internet and how to compile and link your programs, refer to Appendix B.

The following example displays a window with physical dimensions of 320 by 240 pixels and a background color of red. We set the clipping area (or the world coordinates) to start from (0,0) and extend to (160,120). The viewport is set to occupy the entire window, or 320 by 240 pixels. This setting means that every increment of one coordinate in our application will be mapped to two pixel increments in the physical coordinate system (application window boundaries (0,0) to (160,120) vs. physical window boundaries (0,0) to (320,240)). If the viewport had been defined to be only 160 by 120 pixels, then there would have been a one-to-one mapping from points in the world coordinate space to the physical pixels.

However, only one fourth of the window would have been occupied by the application! Depending on where you install the example files, you can find the source code for this example in: *Example1_1/Example1_1.cpp.*

```
//Example1_1.cpp: A simple example to open a window
// the windows include file, required by all windows apps
#include <windows.h>

// the glut file for windows operations
// it also includes gl.h and glu.h for the OpenGL library calls
#include <gl\glut.h>

void Display(void)
{
        //clear all pixels with the specified clear color
        glClear(GL_COLOR_BUFFER_BIT);
//don't wait, start flushing OpenGL calls to display buffer
        glFlush();
}
void init(void){
        //set the clear color to be red
        glClearColor(1.0,0.0,0.0,1.0);
        //set the viewport to be 320 by 240, the initial size of the window
        glViewport(0,0,320,240);
// set the 2D clipping area
        gluOrtho2D(0.0, 160.0, 0.0, 120.0);
}

void main(int argc, char* argv[])
{
        glutInitDisplayMode(GLUT_SINGLE | GLUT_RGB);
```

```
        glutInitWindowSize (320, 240);
        glutCreateWindow("My First OpenGL Window");
        init();
        glutDisplayFunc(Display);
        glutMainLoop();
}
```

Since this is our first OpenGL program, Let us understand the example line by line.

The Include Files

There are only two include files:

```
#include <windows.h>
#include <gl\glut.h>
```

The windows.h is required by all windows applications. The header file glut.h includes the GLUT library functions as well as gl.h and glu.h, the header files for the OpenGL library functions. All calls to the glut library functions are prefixed with glut. Similarly, all calls to the OpenGL library functions start with the prefix gl or glu.

The Body

Let us look at the main program first. The line

```
glutInitDisplayMode(GLUT_SINGLE | GLUT_RGB);
```

tells the GLUT library what type of display mode to use when creating the application window. In this case, we specify that we will be using only a single frame buffer (GLUT_SINGLE) and we want to specify colors in the RGB color mode (GLUT_RGB). We will discuss more about frame buffers in a later chapter.
The next call

```
glutInitWindowSize (320, 240);
```

initializes the window to have an initial size (physical resolution) of 320 by 240 pixels.
The call

```
glutCreateWindow("My First OpenGL Window");
```

actually creates the window with the caption "My First OpenGL Window".

The next function

```
init();
```

initializes some of the OpenGL parameters before we actually display the rendered window.

OpenGL and glut work with the help of callback functions. Events that occur on the computer (such as mouse clicks, keyboard clicks, moving the window etc.) that you wish your program to react to need to be registered with OpenGL as callback functions. When the event occurs, OpenGL automatically calls the function registered to react to the event appropriately.
The function

```
glutDisplayFunc(Display);
```

registers the callback function for display redrawing to be the function *Display*. The Display callback is triggered whenever the window needs to be redrawn, and GLUT will automatically call the Display function for you. When does the window need to be redrawn? When you first display the window, when you resize the window, or even when you move the window around. We shall see what the init function and the Display function actually do in a bit.
Finally, we make a call to the glut library function

```
glutMainLoop();
```

This function simply loops, monitoring user actions and making the necessary calls to the specified callback functions (in this case, the '*Display*' function) until the program is terminated.

The init() function

The init function itself is defined to initialize the GL environment. It does this by making three calls:

```
glClearColor(1.0, 0.0, 0.0, 1.0);
```

This gl library command sets the color for clearing out the contents in the frame buffer (which then get drawn into the window). It expects the RGB values, in that order, as parameter,s as well as the alpha component of the color. For now, we set the alpha to always be 1. The above command will set the clear color to be pure red. Try experimenting with different clear colors and see what effect this has on the window display.

Next, we define the viewport to be equal to the initial size of the window by calling the function

```
glViewport(0, 0, 320, 240)
```

And we set the clipping area, or our world coordinate system, to be (0,0) to (160,120) with the glu library command

```
gluOrtho2D(0.0, 160.0, 0.0, 120.0);
```

The Display() function

The Display function simply makes two OpenGL calls:

```
glClear(GL_COLOR_BUFFER_BIT);
```

On a computer, the memory (frame buffer) holding the picture is usually filled with the last picture you drew, so you typically need to clear it with some background color before you start to draw the new scene.

OpenGL provides glClear as a special command to clear a window. This command can be much more efficient than a general-purpose drawing command since it clears the entire frame buffer to the current clearing color. In the present example, we have set the clear color earlier to be red.

In OpenGL, the frame buffer can be further broken down into buffers that hold specialized information.

The color buffer (defined as GL_COLOR_BUFFER_BIT) holds the color information for the pixel. Later on, we shall see how the depth buffer holds depth information for each pixel.

The single parameter to glClear() indicates which buffers are to be cleared. In this case, the program clears only the color buffer.
Finally, the function

```
glFlush();
```

forces all previously issued OpenGL commands to begin execution. If you are writing your program to execute within a single machine, and all commands are truly executed immediately on the server, glFlush() might have no effect. However, if you're writing a program that you want to work properly both with and without a network, include a call to glFlush() at the end of each frame or scene. Note that glFlush() doesn't wait for the drawing to complete —it just forces the drawing to begin execution.

Voila: when you run the program you will see a red window with the caption "My First OpenGL window". The program may not seem very interesting, but it demonstrates the basics of getting a window up and running using OpenGL. Now that we know how to open a window, we are ready to start drawing into it.

Plotting Points

Objects and scenes that you create in computer graphics usually consist of a combination of shapes arranged in unique combinations. Basic shapes, such as points, lines, circles, etc., are known as graphics primitives. The most basic primitive shape is a point.

In the previous section, we considered Cartesian coordinates and how we can map the world coordinate system to actual physical screen coordinates. Any point that we define in our world has to be mapped onto the actual physical screen coordinates in order for the correct pixel to light up. Luckily, for us, OpenGL handles all this mapping for us; we just have to work within the extent of our defined world coordinate system. A point is represented in OpenGL by a set of floating-point numbers and is called a *vertex*.

We can draw a point in our window by making a call to the gl library function

```
glVertex{2,3,4}{s,I,d,f}
```

The {2,3,4} option indicates how many coordinates define the vertex and the {s,i,d,f} option defines whether the arguments are short, integers, double precision, or floating-point values. By default, all integer values are internally converted to floating-point values.
For example, a call to

```
glVertext2f(1.0,2.0)
```

refers to a vertex point (in world coordinate space) at coordinates (1.0,2.0). Almost all library functions in OpenGL use this format. Unless otherwise stated, we will always use the floating-point version of all functions. To tell OpenGL what set of primitives you want to define with the vertices, you bracket each set of vertices between a call to

glBegin() and **glEnd()**.

The argument passed to glBegin() determines what sort of geometric primitive it is. To draw vertex points, the primitive used is GL_POINTS. We modify *Example1_1* to draw four points. Each point is at a distance of (10,10) coordinates away from the corners of the window and is drawn with a different color. Compile and execute the code shown below. You can also find the code for this example under *Example1_2/Example1_2.cpp*..

```
// Example1_2.cpp: let the drawing begin

#include <windows.h>
#include <gl\glut.h>

void Display(void)

{
```

```
        //clear all pixels with the specified clear color
        glClear(GL_COLOR_BUFFER_BIT);
        //draw the four points in four colors
        glBegin(GL_POINTS);
                glColor3f(0.0, 1.0, 0.0);               // green
                glVertex2f(10.,10.);
                glColor3f(1.0, 1.0, 0.0);               // yellow
                glVertex2f(10.,110.);
                glColor3f(0.0, 0.0, 1.0);               // blue
                glVertex2f(150.,110.);
                glColor3f(1.0, 1.0, 1.0);               // white
                glVertex2f(150.,10.);
        glEnd();

        //dont wait, start flushing OpenGL calls to display buffer
        glFlush();
}

void reshape (int w, int h)
{
        // on reshape and on startup, keep the viewport to be the entire size of the window
        glViewport (0, 0, (GLsizei) w, (GLsizei) h);
        glMatrixMode (GL_PROJECTION);
        glLoadIdentity ();
        // keep our world coordinate system constant
        gluOrtho2D(0.0, 160.0, 0.0, 120.0);
}

void init(void){
        glClearColor(1.0,0.0,0.0,1.0);
        // set the point size to be 5.0 pixels
        glPointSize(5.0);
}

void main(int argc, char* argv[])
{
        glutInitDisplayMode(GLUT_SINGLE | GLUT_RGB);
        glutInitWindowSize (320, 240);
        glutCreateWindow("My First OpenGL Window");
        init();
        glutDisplayFunc(Display);
        glutReshapeFunc(reshape);
        glutMainLoop();
}
```

Most of the code should be self-explanatory.
The main function sets up the initial OpenGL environment.
In the init() function, we set the point size of each vertex drawn to be 5 pixels, by calling the function

```
glPointSize(3.0);
```

The parameter defines the size in pixels of the points being drawn. A 5 pixel point is large enough for us to see it without squinting our eyes too much!
In this function, we also we set the clear color to be red. The Display function defines all the drawing routines needed
The function call

```
glColor3f(0.0, 1.0, 0.0);                    // green
```

sets the color for the next openGL call. The parameters are, in order, the red, green, and blue components of the color. In this case, we redefine the color before plotting every vertex point. We define the actual vertex points by calling the function

```
glVertex2f(10.,10.);
```

with the appropriate *(x,y)* coordinates of the point. In this example, we define two callback functions. We saw how to define the callback function for redrawing the window. For the rest of the book, we will stick with the convention of this function being called "Display".

In this example we define the viewport and clipping area settings in a new callback function called *reshape*. We register this callback function with OpenGL with the command

```
glutReshapeFunc(reshape);
```

This means that the reshape function will be called whenever the window resizes itself (which includes the first time it is drawn on the screen!) The function receives the width and height of the newly shaped window as its arguments. Every time the window is resized, we reset the viewport so as to always cover the entire window. We always define the world coordinate system to remain constant at ((0,0),(160,120)). As you resize the window, you will see that the points retain their distance from the corners. What mapping is being defined? If we change the clipping area to be defined as

```
gluOrtho2D(0.0, (Gldouble)w, 0.0, (Gldouble)h);
```

you would see that the points maintain their distance from each other and not from the corners of the window. Why?

1.4 Shapes and Scan Converting

We are all familiar with basic shapes such as lines and polygons. They are easy enough to visualize and represent on paper. But how do we draw them on the computer? The trick is in finding the right pixels to turn on!

The process by which an idealized shape, such as a line or a circle, is transformed into the correct "on" values for a group of pixels on the computer is called scan conversion and is also referred to as *rasterizing*.

Over the years, several algorithms have been devised to make the process of scan converting basic geometric entities such as lines and circles simple and fast.

The most popular line-drawing algorithm is the midpoint-line algorithm. This algorithm takes the *x*- and *y*- coordinates of a line's endpoints as input and then calculates the *x,y*-coordinate pairs of all the pixels in between. The algorithm begins by first calculating the physical pixels for each endpoint. An ideal line is then drawn connecting the end pixels and is used as a reference to determine which pixels to light up along the way. Pixels that lie less than 0.5 units from the line are turned on, resulting in the pixel illumination as shown in Fig.1.12.

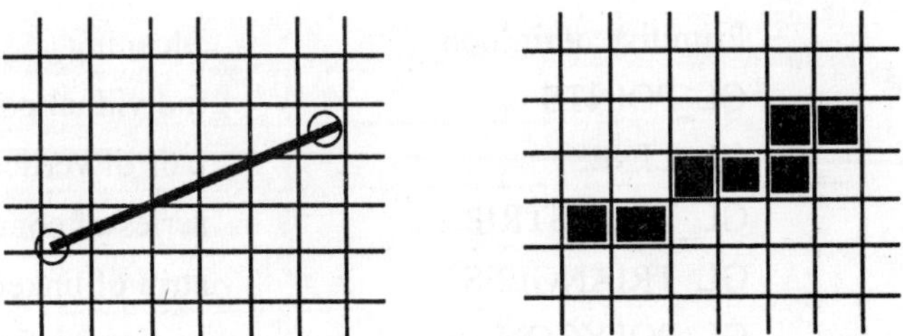

Fig.1.12: Midpoint algorithm for line drawing

All graphic packages (OpenGL included) incorporate predefined algorithms to calculate the pixel illuminations for drawing lines.

Basic linear shapes such as triangles and polygons can be defined by a series of lines. A polygon is defined by *n* number of vertices connected by lines, where n is the number of sides in the polygon. A quadrilateral, which is a special case of a polygon is defined by four vertices, and a triangle is a polygon with three vertices as shown in Fig.1.13.

To specify the vertices of these shapes in OpenGL, we use the function that we saw earlier:

glVertex.

To tell OpenGL what shape you want to create with the specified vertices, you bracket each set of vertices between a call to glBegin() and a call to glEnd(). The

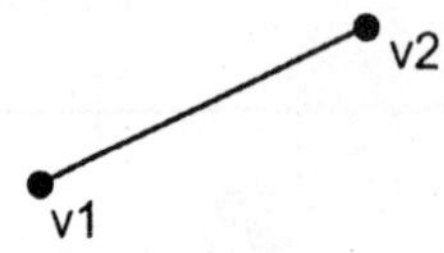

1) A line is defined by 2 vertices

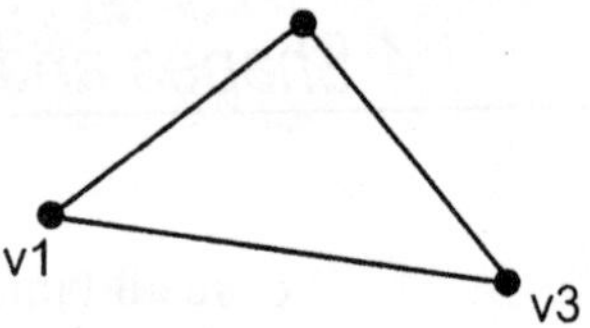

2) A triangle is defined by 3 vertices

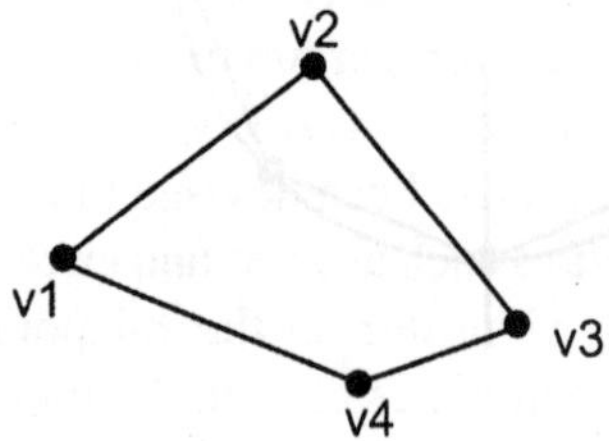

3) A quadrilateral is defined by 4 verticles

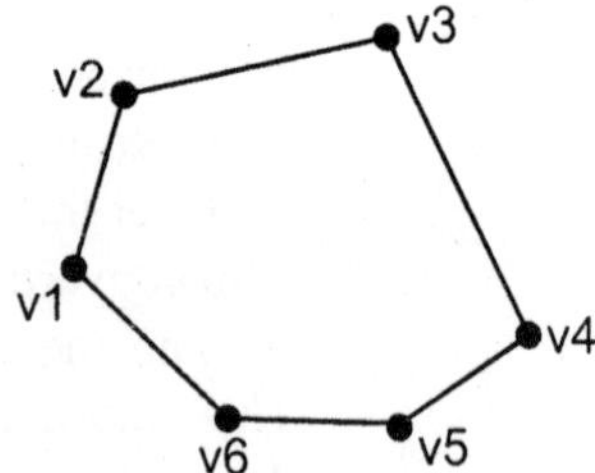

4) An n-polygon is defined by n vertices

Fig.1.13: Vertices needed to define different kinds of basic shapes

argument passed to glBegin() determines what sort of geometric primitive; some of the commonly used ones are described in Table 1.2.

Primitive definition	Meaning
GL_POINTS	individual points
GL_LINES	pair of vertices defining a line
GL_LINE_STRIP	series of connected lines
GL_TRIANGLES	strip of linked triangles
GL_POLYGON	vertices define a simple convex polygon

Table 1.2: OpenGL geometric primitive types

Note that primitives are all straight-line primitives. There are algorithms like the midpoint algorithm that can scan convert shapes like circles and other hyperbolic fig.s. The basic mechanics for these algorithms are the same as for lines: figure out the pixels along the path of the shape, and turn the appropriate pixels on.

Interestingly, we can also draw a curved segment by approximating its shape using line segments. The smaller the segments, the closer the approximation.

For example, consider a circle. Recall from trigonometry that any point on a circle of radius *r* (and centered at the origin) has an *x,y*-coordinate pair that can be represented as a function of the angle theta the point makes with the axes, as shown in Fig.1.14.

$$P(\theta)=((r\ \cos\theta),(r\ \sin\theta))$$

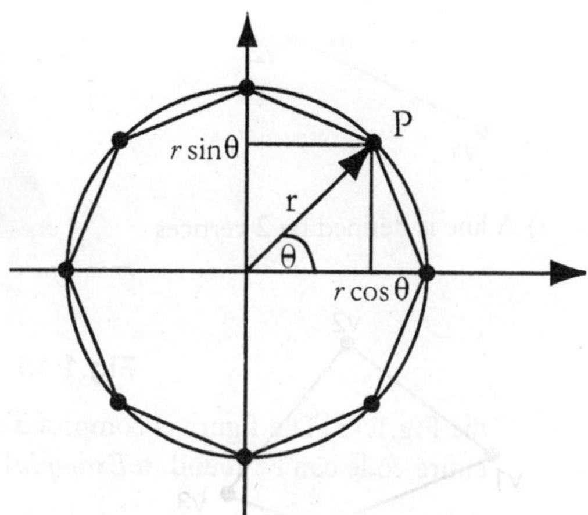

Fig.1.14: Points along a circle

As we vary theta from 0 to 360 degrees (one whole circle), we can get the (*x,y*) coordinates of points along the circle. So if we can just plot "enough" of these points and draw line segments between them, we should get a fig. that looks close enough to a circle. A sample code snippet to draw a circle with approximately 100 points is shown below. Note that we need to add the center of the circle to our equations to position our circle appropriately.

```
#define PI 3.1415926535898
// cos and sin functions require angles in radians
// recall that 2PI radians = 360 degrees, a full circle

GLint circle_points = 100;
void MyCircle2f(GLfloat centerx, GLfloat centery, GLfloat radius){
        GLint i;
        GLdouble theta;
        glBegin(GL_POLYGON);
        for (i = 0; i < circle_points; i++) {
                theta= 2*PI*i/circle_points; // angle in radians
                glVertex2f(centerx+radius*cos(theta),
                        centery +radius*sin(theta));
        }
        glEnd();
}
```

Remember that the math functions *cos* and *sin* require angles in radians and that 2PI radians make 360 degrees—hence our conversions in the code. We can construct more complex shapes by putting these basic primitives together. Shown below is a snippet of code to draw a stick figure of a person as shown in

Fig.1.15: Stick Figure

the Fig.1.15. The figure is composed of lines, polygons, points and a circle. The entire code can be found in *Example1_3/Example1_3.cpp*.

```
// Example 1_3.cpp

void Display(void)
{
//clear all pixels with the specified clear color
        glClear(GL_COLOR_BUFFER_BIT);
        glColor3f(1.0,0.8,0.1);
        MyCircle2f(80.,85., 10.);

// the eyes are black points
// set the point size to be 3.0 pixels
        glBegin(GL_POINTS);
                glColor3f(0.0, 0.0, 0.0);
                glVertex2f(77.,88.);
                glVertex2f(83.,88.);
        glEnd();

// polygonal body
        glColor3f(0.8,0.0,0.9);
        glBegin(GL_POLYGON);
                glVertex2f(75.,75.);
                glVertex2f(85.,75.);
                glVertex2f(100.,30.);
                glVertex2f(60.,30.);
        glEnd();

//rectangular legs
        glColor3f(1.0,0.8,0.1);
        glRectf(70.,5.,75.,30.);
        glRectf(85.,5.,90.,30.);
```

```
// but lines for hands!
        glBegin(GL_LINES);
                glVertex2f (74.,70.); glVertex2f (50.,50.);
        glEnd();
        glBegin(GL_LINES);
                glVertex2f (86.,70.); glVertex2f (110.,50.);
        glEnd();

//don't wait, start flushing OpenGL calls to display buffer
        glFlush();
}
```

Note that with OpenGL, the description of the shape of an object being drawn is independent of the description of its color. Whenever a particular geometric object is drawn, it is drawn using the currently specified coloring scheme. Until the color or coloring scheme is changed, all objects are drawn in the current coloring scheme. Similarly, until the point or line sizes are changed, all such primitives will be drawn using the most currently specified size. Try composing your own objects by putting smaller primitives together.

If you run the above program, you may notice that the slanting lines appear to be jagged. This is an artifact caused due to the algorithms that we employ to rasterize the shapes and is known as *aliasing*.

Anti-Aliasing

The problem with most of the scan conversion routines is that the conversion is jagged. This effect is an unfortunate consequence of our all or nothing approach to illuminating pixels. At high resolutions, where pixels are close together, thi effect is not noticeable, but on low-resolution monitors, it produces a harsh, jagged look. Aliasing can be a huge problem in computer generated movies, when you can sometimes actually see jagged lines crawling from scene to scene, creating a disturbing effect.

A solution to this problem is called *anti-aliasing*. It employs the principle that if pixels are set to different intensities, and if adjoining pixel intensities can be properly manipulated, then the pixels will blend to form a smooth image. So going back to the midpoint algorithm, as shown in Fig.1.16, anti-aliasing would turn pixels on with varying intensities (depending on how a one-unit thick line would intersect with the pixels), instead of merely turning pixels on and off. This process tends to make the image look blurry but more continuous.

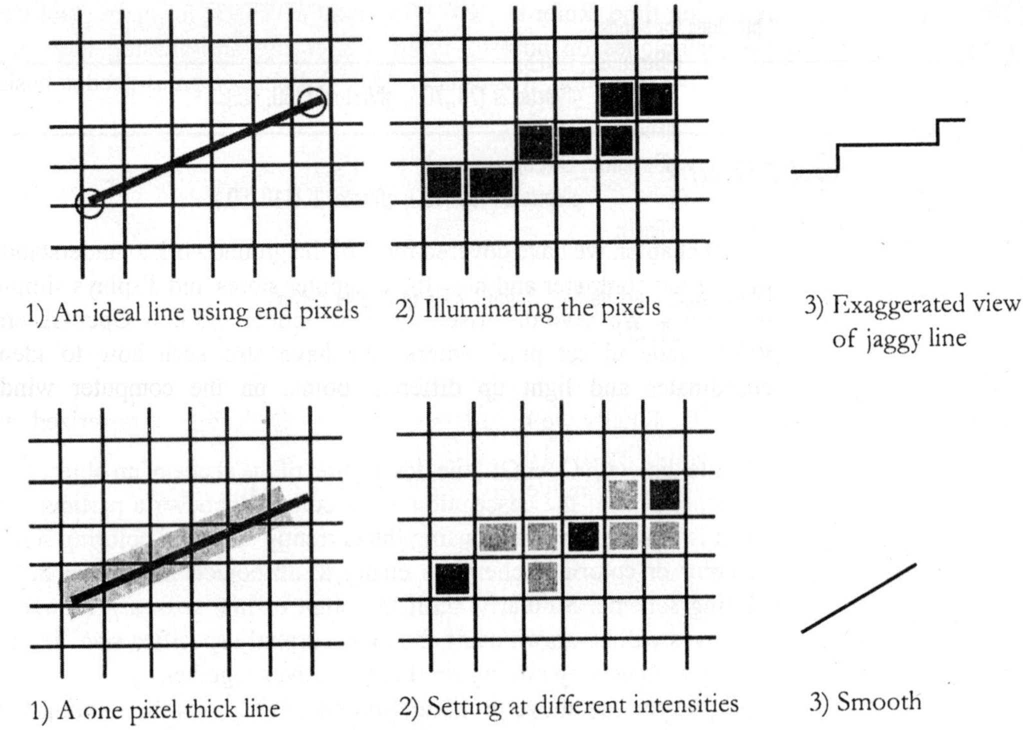

Fig.1.16: Anti-aliasing a line

To deploy anti-aliasing in OpenGL, there are two steps we need to take:

1. To antialias points or lines, you need to turn on antialiasing with glEnable(), passing in GL_POINT_SMOOTH or GL_LINE_SMOOTH, as appropriate.
2. We also need to enable blending by using a blending factor. The blending factors you most likely want to use are GL_SRC_ALPHA (source) and GL_ONE_MINUS_SRC_ALPHA (destination).

To anti-alias our stick figure, we add a few more calls in the init function from *Example1_3*, as shown in the code below:

```
glEnable (GL_LINE_SMOOTH);
glEnable (GL_BLEND);
  glBlendFunc(GL_SRC_ALPHA,GL_ONE_MINUS_SRC_ALPHA);
glHint(GL_LINE_SMOOTH_HINT | GL_POLYGON_SMOOTH_HINT, GL_DONT_CARE);
```

You can find the entire source code in *Example1_4/Example1_4.cpp*. When you run this example, look at the hands carefully. You will notice they seem smoother than from *Example1_3*. (Note that the polygons are still not anti-aliased as they need further treatment.) The actual details of calculating the intensities of different pixels can be complicated and usually results in slower rendering time. Refer to [FOLE95] and [WATT93] for more details on aliasing and techniques on how to avoid it. Because anti-aliasing is a complex and expensive operation, it is usually deployed only on an as-needed basis.

Summary

In this chapter, we have covered most of the groundwork to understand the workings of the computer and how the computer stores and displays simple graphics primitives. We have discussed the color model and how OpenGL employs the RGB mode to set pixel colors. We have also seen how to identify pixel coordinates and light up different points on the computer window using OpenGL. Finally we have learned how basic shapes are rasterized, anti-aliased and finally displayed on a grid of pixels. In the next chapter, we will explore how to construct and move the shapes around in our 2D world.

Summary

Chapter 2

Making Them Move

In the previous chapter, we saw how to draw basic shapes using the OpenGL graphics library. But we want to be able to do more than just draw shapes: we want to be able to move them around to design and compose our 2D scene. We want to be able to scale the objects to different sizes and orient them differently. The functions used for modifying objects such as translation, rotation, and scaling are called *geometric transformations*.

Why do we care about transformations? Usually we define our shapes in a coordinate system convenient to us. Using transformation equations enables us to easily modify object locations in the world coordinate system. In addition, if we transform our objects and play back the CG images fast enough, our eyes will be tricked to believe we are seeing motion. This principle is used extensively in animation, and we shall look into it in detail in Chapter 7. When we study 3D models, we will also see how we can use transformations to define hierarchical objects and to "transform" the camera position.

This chapter introduces the basic 2D transformations used in CG: translation, rotation, and scaling. These transformations are essential ingredients of all graphics applications.

In this chapter you will learn the following concepts:

- Vectors and matrices
- 2D Transformations: translation, scaling, and rotation
- How to use OpenGL to transform objects
- Composition of transforms

2.1 Vectors and Matrices

Before we jump into the fairly mathematical discussion of transformations, let us first brush up on the basics of vector and matrix math. This math will form the basis for the transformation equations we shall see later in this chapter. We discuss the math involved as applied to a 2D space. The principles are easily extended to 3D by simply adding a third axes, namely, the *z*- axis.

Vectors

A vector is a quantity that has both direction and length. In CG, a vector represents a directed line segment with a start point (its tail) and an end point (the head, shown typically as an arrow pointed along the direction of the vector). The length of the line is the length of the vector.

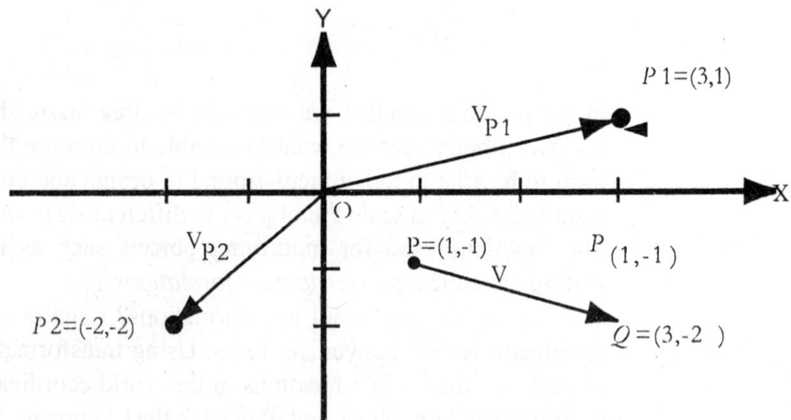

Fig. 2.1: A 2D Vector/Point: A directional line

A 2D vector that has a length of *x* units along the *x*-axis and *y* units along the *y*-axis is denoted as $\begin{bmatrix} x \\ y \end{bmatrix}$

It is valuable to think of a vector as a displacement from one point to another. Consider two points *P*(1,-1) and *Q*(3,-2), as shown in Figure 2.1. The displacement from *P* to *Q* is the vector $\mathbf{V} = \begin{bmatrix} 2 \\ -1 \end{bmatrix}$, calculated by subtracting the coordinates of the points individually. What this means is that to get from *P* to *Q*, we shift right along the *x*-axis by two units and down the *y*-axis by one unit. Interestingly, any point *P*1 with coordinates (x,y) corresponds to the vector $\mathbf{V}_{P1}$, with its head at (*x*,*y*) and tail at (0,0) as shown in Figure 2.1. That is, there is a one-to-one correspondence between a vector, with its tail at the origin, and a point on the plane. This means that we can also represent the point P(x,y) by the vector $\begin{bmatrix} x \\ y \end{bmatrix}$

Often, the math of transformation equations uses the vector representation of points in this manner, so do not let this usage confuse you.

Operations with Vectors

Vectors support some fundamental operations: addition, subtraction, and multiplication with a real number.

Vectors can be added by performing componentwise addition. If V_1 is the vector $(x1,y1)$ and V_2 is the vector $(x2,y2)$ then V_1+V_2 is

$$\begin{bmatrix} x1 \\ y1 \end{bmatrix} + \begin{bmatrix} x2 \\ y2 \end{bmatrix} = \begin{bmatrix} x1 + x2 \\ y1 + y2 \end{bmatrix}$$

Conceptually, adding two vectors results in a third vector which is the addition of one displacement with another.

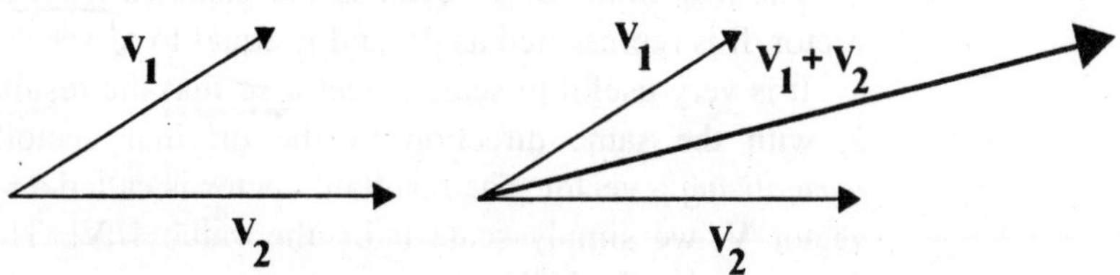

Fig. 2.2: Adding two vectors

Multiplying a vector by a number s results in a vector whose length has been scaled by s. For this reason, the number s is also referred to as a scalar. If s is negative, then this results in a vector whose direction is flipped as well.

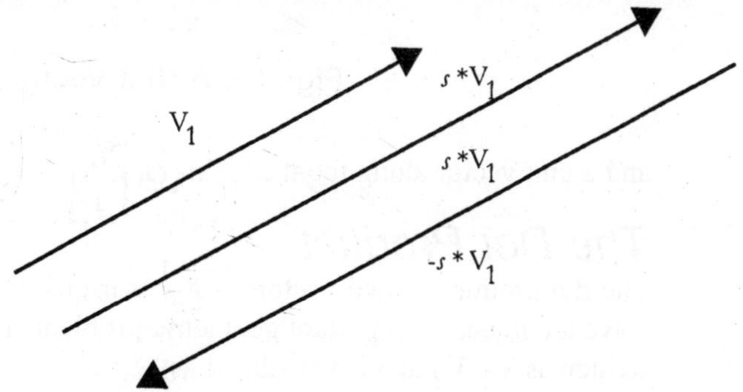

Fig. 2.3: Multiplying a vector with a scalar

Mathematically, the scaled vector $sV_1 = \begin{bmatrix} s*x1 \\ s*y1 \end{bmatrix}$

Subtraction follows easily as the addition of a vector that has been flipped: that is $\mathbf{V_1} - \mathbf{V_2} = \mathbf{V_1} + (-\mathbf{V_2})$.

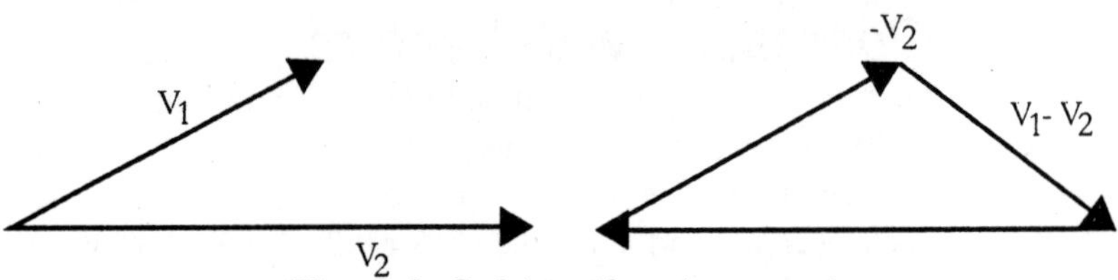

Fig. 2.4: Subtracting two vectors

The Magnitude of a Vector

The length of a vector $\mathbf{V} = \begin{bmatrix} x \\ y \end{bmatrix}$ is also referred to as its *magnitude*.

The magnitude of a vector is the distance from the tail to the head of the vector. It is represented as $|\mathbf{V}|$ and is equal to $\sqrt{x^2+y^2}$

It is very useful to scale a vector so that the resultant vector has a length of 1, with the same direction as the original vector. This process is called *normalizing* a vector. The resultant vector is called a *unit vector*. To normalize a vector **V**, we simply scale it by the value $1/|\mathbf{V}|$. The resultant unit vector is represented as $\mathbf{V} = \mathbf{V}/|\mathbf{V}|$.

Interestingly, a unit vector along the x-axis is quite simply the vector $\begin{bmatrix} 1 \\ 0 \end{bmatrix}$

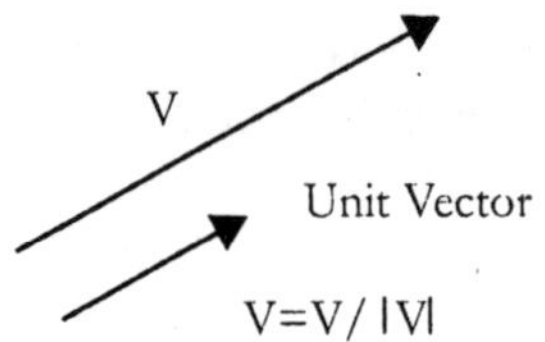

Fig. 2.5: A Unit Vector

and a unit vector along the the y axis is $\begin{bmatrix} 0 \\ 1 \end{bmatrix}$

The Dot Product

The dot product of two vectors is a scalar quantity. The dot product is used to solve a number of important geometric problems in graphics. The dot product is written as $\mathbf{V_1}.\mathbf{V_2}$ and is calculated as

$$\mathbf{V_1}.\mathbf{V_2} = \begin{bmatrix} x1 \\ y1 \end{bmatrix} \bullet \begin{bmatrix} x2 \\ y2 \end{bmatrix} = x1 * x2 + y1 * y2$$

You can find many of these vector functions coded in a utility include file provided by us called *utils.h.*

Matrices

A matrix is an array of numbers. The number of rows (*m*) and columns (*n*) in the array defines the cardinality of the matrix (*m* x *n*). In reality, a vector is simply a matrix with one column (or a one-dimensional matrix). We saw how displays are just a matrix of pixels. A frame buffer is a matrix of pixel values for each pixel on the display.

Matrices can be added component wise, provided they have the same cardinality:

$$\begin{bmatrix} a11 & a12 \\ a21 & a22 \end{bmatrix} + \begin{bmatrix} b11 & b12 \\ b21 & b22 \end{bmatrix} = \begin{bmatrix} a11+b11 & a12+b12 \\ a21+b21 & a22+b22 \end{bmatrix}$$

Multiplication of a matrix by a scalar is simply the multiplication of its components by this scalar:

$$s^*\begin{bmatrix} a11 & a12 \\ a21 & a22 \end{bmatrix} = \begin{bmatrix} s^*a11 & s^*a12 \\ s^*a21 & s^*a22 \end{bmatrix}$$

Two matrices can be multiplied if and only if the number of columns of the first matrix (*m* x *n*) is equal to the number of rows of the second (*n* x *p*). The result is calculated by applying the dot product of each row of the first matrix with each column of the second. The resultant matrix has a cardinality of (*m* x *p*).

$$\begin{bmatrix} a11 & a12 & a13 \\ a21 & a22 & a23 \end{bmatrix} * \begin{bmatrix} b11 & b12 \\ b21 & b22 \\ b31 & b32 \end{bmatrix} =$$

$$\begin{bmatrix} a11*b11 + a12*b21 + a12*b31 & a11*b12 + a12*b22 + a13*b32 \\ a21*b11 + a22*b21 + a23*b31 & a21*b12 + a22*b22 + a23*b32 \end{bmatrix}$$

We can multiply a vector by a matrix if and only if it satisfies the rule above.

$$\begin{bmatrix} a11 & a12 & a13 \\ a21 & a22 & a23 \\ a31 & a32 & a33 \end{bmatrix} * \begin{bmatrix} v1 \\ v2 \\ v3 \end{bmatrix} = \begin{bmatrix} a11*v1 + a12*v2 + a13*v3 \\ a21*v2 + a22*v2 + a23*v2 \\ a31*v3 + a32*v3 + a33*v3 \end{bmatrix}$$

An *identity matrix* is a square matrix (an equal number of rows and columns) with all zeroes, except for 1s in its diagonal.

$$\begin{bmatrix} 1 & 0 & 0 \\ 0 & 1 & 0 \\ 0 & 0 & 1 \end{bmatrix}$$

Multiplication of a vector/matrix by an identity matrix has no effect. Prove this by multiplying a vector with the appropriate identity matrix. Refer to LENG93 for more details on vectors and matrices.

Whew! After that fairly exhaustive review of matrices and vectors, we are ready to get into the thick of transformations.

2.2 2D Object Transformations

The functions used for modifying the size, location, and orientation of objects or of the camera in a CG world are called *geometric transformations*. Transformations are applied within a particular space domain, be it an object space or the camera space or a texture space. We shall mainly be discussing transformations applied in the object space, also called *object transformations or model transformations*. The mathematics for transformations in other spaces remains the same.

Object Space

Usually, objects are defined in a default coordinate system convenient to the modeler. This coordinate system is called the object coordinate system or object space.

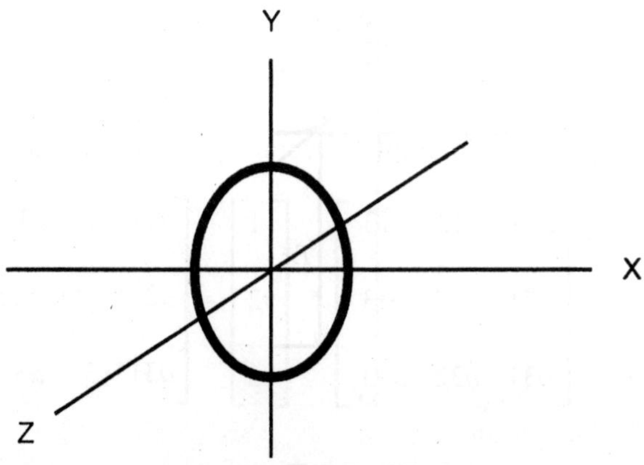

Fig.2.5: Object space

The object coordinate system can be transformed, in order to transform the object within the world coordinate system. An object transformation sets the state of the object space. Internally, this state is represented as a matrix. All the objects drawn on the screen after this state is set are drawn with the new transformations applied. Transformations make it convenient to move objects within the world, without actually having to model the object again!

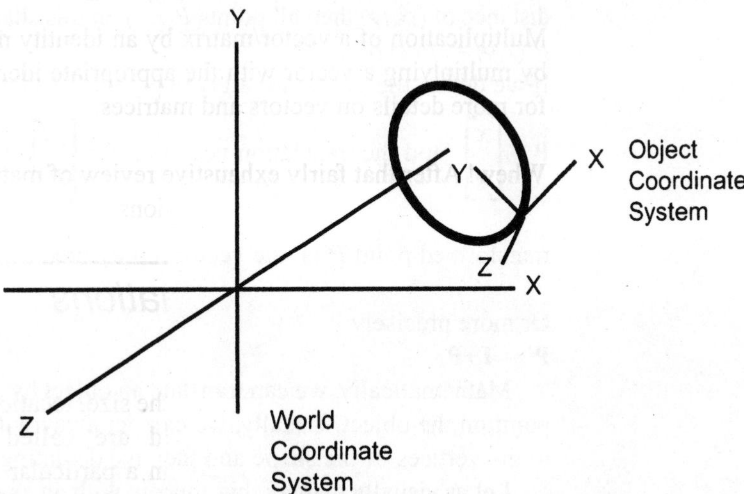

Fig.2.6: Object space is transformed within the world coordinate system.

Let us look into the three kinds of transformations most commonly used in CG: translation, rotation, and scaling.

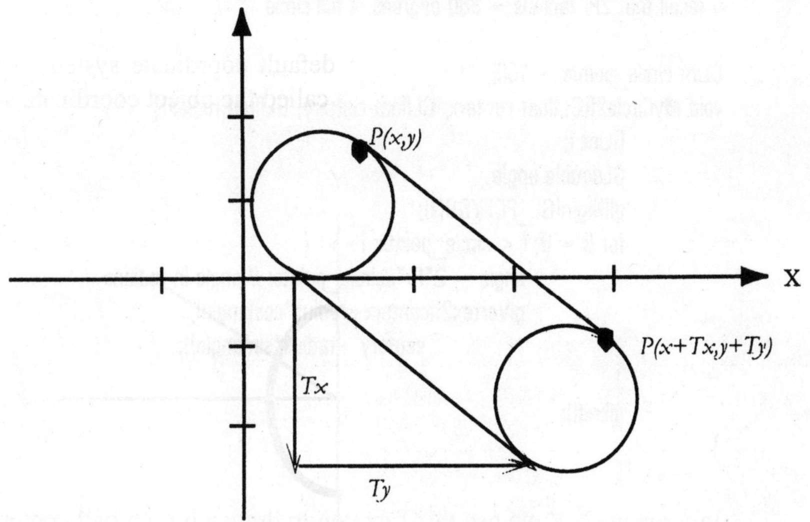

Fig.2.7:Translation along the x- and y-axes

Translation

Translation is the transform applied when we wish to move an object. In a 2D world, we can move the object from left to right (translate along the x-axis) or up and down (translate along the y-axis) as shown in Fig.2.7. The abbreviations for these translations are Tx and Ty.

Consider a circular shape, P, as shown in Figure. If we wish to move P by a distance of (tx,ty) then all points $P(x,y)$ on this shape will move to a new location $P'(x,y) = P(x+tx, y+ty)$.

If we define the vector for a given points $P(x,y)$ as:

$\mathbf{P} = \begin{bmatrix} x \\ y \end{bmatrix}$ and the translation vector as $\mathbf{T} = \begin{bmatrix} Tx \\ Ty \end{bmatrix}$ then the resultant transformed point P' is the vector represented as $\mathbf{P'} = \begin{bmatrix} x + Tx \\ y + Ty \end{bmatrix}$

Or more precisely

$$\mathbf{P'} = \mathbf{T} + \mathbf{P}$$

Mathematically, we can translate an object by applying this equation to every point on the object. Usually, we can get away with just applying the translation to the vertices of the shape and then recalculating the translated shape.

Let us visually display this concept with an example of a bouncing ball. From the last chapter, you may recall how to draw a circle:

```
#define PI 3.1415926535898
// cos and sin functions require angles in radians
// recall that 2PI radians - 360 degrees, a full circle

GLint circle_points - 100;
void MyCircle2f(GLfloat centerx, GLfloat centery, GLfloat radius){
        GLint i;
        GLdouble angle;
        glBegin(GL_POLYGON);
        for (i - 0; i < circle_points; i++) {
                angle - 2*PI*i/circle_points; // angle in radians
                glVertex2f(centerx+radius*cos(angle),
                        centery +radius*sin(angle));
        }
        glEnd();
}
```

In *Example2_1*, we use this function to draw a brown ball centered at the origin with the radius of the ball set to be 15 units in our world coordinate system:

```
GLfloat RadiusOfBall = 15.;
// Draw the ball, centered at the origin
void draw_ball() {
  glColor3f(0.6,0.3,0.);
  MyCircle2f(0.,0.,RadiusOfBall);
}
```

We want to bounce this ball up and down in our display window. That is, we wish to translate it along the *y*-axis. The floating-point routine for performing translations in OpenGL is

```
glTranslatef(Tx, Ty, Tz)
```

It accepts three arguments for translation along the *x*-, *y*-, and *z*-axes respectively. The *z*-axis is used for 3D worlds, and we can ignore it for now by setting the *Tz* value to be 0.

Before we apply a new translation, we need to set the transformation state in the object space to be an identity matrix (i.e., no transformation is being applied). The command

```
glLoadIdentity ()
```

clears out the current transformation state with the identity matrix. We do this because we do not wish to reuse old transformation states.

Then, we constantly add (or subtract) to the *ty* component along the *y*-axis, and draw the ball in this newly transformed position. The snippet of code to translate an object in a window with the maximum extent of the world coordinates set to (160,120) is shown below. A *ydir* variable is used to define whether the ball is bouncing up or down (and hence whether we should increment or decrement *ty*).

```
// 160 is max X value in our world
// Define X position of the ball to be at center of window
xpos = 80.;
// 120 is max Y value in our world
// set Y position to increment 1.5 times the direction of the bounce
  ypos = ypos+ydir *1.5;
// If ball touches the top, change direction of ball downwards
  if (ypos == 120-RadiusOfBall)
          ydir = -1;
// If ball touches the bottom, change direction of ball downwards
  else if (ypos <RadiusOfBall)
          ydir = 1;
```

```
//reset transformation state
glLoadIdentity();
// apply the desired translation
glTranslatef(xpos,ypos, 0.);
```

Camera (or viewing) and object transformations are combined into a single matrix in OpenGL, called the model-view matrix (GL_MODELVIEW). The command

```
glMatrixMode(GL_MODELVIEW)
```

specifies the space where the transformations are being applied. The mode is normally set in the reshape function where we also set the clipping area of the window.

That is, whenever the window is reshaped, resized, or first drawn, we specify any further transformations to be applied in the object space.

If you try to view this animation, we shall see motion that is jerky, and you may even see incomplete images being drawn on the screen. To make this bouncing motion appear smooth, we make use of a concept called *double buffering.*

Double Buffering

Double buffering provides two frame buffers for use in drawing. One buffer, called the foreground buffer is used to display on the screen. The other buffer, called the background buffer, is used to draw into. When the drawing is complete, the two buffers are swapped so that the one that was being viewed is now being used for drawing and vice versa. The swap is almost instantaneous. Since the image is already drawn when we display it on screen, it makes the resulting animation look smooth, and we don't see incomplete images. The only change in the required to activate double buffering is to specify the display mode to be **GLUT_DOUBLE.**

```
glutInitDisplayMode(GLUT_DOUBLE | GLUT_RGB);
```

This call is made in the main function of *Example2_1*. By default, in double buffer mode, OpenGL renders all drawing commands into the background buffer. A call to

```
glutSwapBuffers();
```

will cause the two buffers to be swapped. After they are swapped, we need to inform OpenGL to redraw the window (using the new contents of the foreground buffer).

The function

```
glutPostRedisplay()
```

forces a re-draw of the window.

The code required to display the bouncing ball is as follows:

```
void Display(void)
{
        glClear(GL_COLOR_BUFFER_BIT);
        glClear(GL_COLOR_BUFFER_BIT);
        // 160 is max X value in our world
        // Define X position of the ball to be at center of window
        xpos = 80.;
        // 120 is max Y value in our world
        // set Y position to increment 1.5 times the direction of the bounce
        ypos = ypos+ydir *1.5;
        // If ball touches the top, change direction of ball downwards
        if (ypos == 120-RadiusOfBall)
                ydir = -1;
        // If ball touches the bottom, change direction of ball downwards
        else if (ypos <RadiusOfBall)
                ydir = 1;
        //reset transformation state
        glLoadIdentity();
        // apply the translation
        glTranslatef(xpos,ypos, 0.);
        // draw the ball with the current transformation state applied
        draw_ball();
        // swap the buffers
        glutSwapBuffers();
        // force a redraw using the new contents
        glutPostRedisplay();
}
```

The entire code for this example can be found under *Example2_1/Example2_1.cpp*

Scaling

An object can be scaled (stretched or shrunk) along the *x*- and *y*- axis by multiplying all its points by the scale factors *Sx* and *Sy*. All points $P=(x,y)$ on the scaled shape will now become $P' = (x',y')$ such that $x' = Sx.x$, $y' = Sy.y$. In matrix form:

$$\begin{bmatrix} x' \\ y' \end{bmatrix} = \begin{bmatrix} Sx & 0 \\ Sy & 0 \end{bmatrix} \begin{bmatrix} x \\ y \end{bmatrix} \quad \text{or } P'=S.P$$

In Figure 2.8, we show the circular shape being scaled by (1/2,2) and by (2,1/2).

Notice from the figure that scaling changes the bottom (base) position of the shape. This is because the scaling equation we defined occurs around the origin

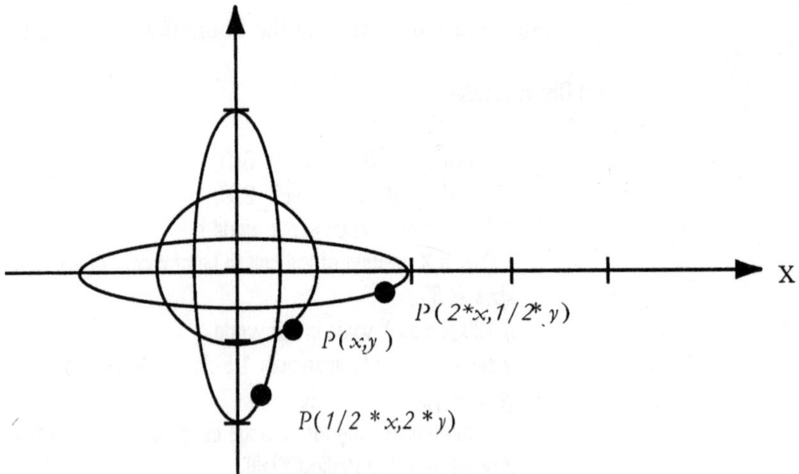

Fig.2.8: Scaling of a circle.

of the world coordinate system. That is, only points at the origin remain unchanged. Many times it is more desirable to scale the shape about some other predefined point. In the case of a bouncing ball, we may prefer to scale about the bottommost point of the ball because we want the base of the shape to remain unchanged. In a later section, we shall see how to scale the shape about a predefined position.

Let us go back to our bouncing ball example. When the ball hits the ground we want it to squash on impact and then stretch back up again into its original shape. This is what we expect from a real (soft) ball bouncing on the ground. The floating point command to scale an object is

```
glScalef(Sx, Sy, Sz)
```

which accepts three arguments for scaling along the *x*-, *y*- and *z*-axes. Scaling has no effect when *Sx*=*Sy*=*Sz*=1. For now, we will set the *Sz* to be 1.

We can modify the Display code from the previous example to include scaling of the ball. When the ball hits the ground, and for some time afterwards, we stop translating and squash the shape. That is, we scale down the shape along the *y*-axis and proportionately scale up along the *x*-axis. When we reach a predetermined squash, we stretch back up again and restart the translation. Shown below is the code to perform this transformation.

```
// Shape has hit the ground! Stop moving and start squashing down and then back up
if (ypos == RadiusOfBall && ydir == -1 ) {
        sy = sy*squash ;
```

```
            if (sy < 0.8)
                    // reached maximum squash, now un-squash back up
                    squash - 1.1;
            else if (sy > 1.) {
                    // reset squash parameters and bounce ball back upwards
                    sy - 1.;
                    squash - 0.9;
                    ydir - 1;
            }
            sx - 1./sy;
}
// 120 is max Y value in our world
// set Y position to increment 1.5 times the direction of the bounce
else {
            ypos - ypos+ydir *1.5 - (1.-sy)*RadiusOfBall;
            if (ypos - - 120-RadiusOfBall)
                    ydir - -1;
            else if (ypos <RadiusOfBall)
                    ydir - 1;
}
glLoadIdentity();
 glTranslatef(xpos,ypos, 0.);
 glScalef(sx,sy, 1.);
 draw_ball();
```

The entire code can be found under *Example 2_2/Example 2_2.cpp*. Notice that two transformations are applied to the object—translation and scaling.

When you run the program, you will notice that the ball doesn't stay on the ground when it squashes! It seems to jump up. This happens because the scaling is happening about the origin—which is the center of the ball!

Rotation

A shape can be rotated about any of the three axes. A rotation about the *z*-axis will actually rotate the shape in the *xy*-plane, which is what we desire in our 2D world. The points are rotated through an angle about the world origin as shown in Fig.2.9. Mathematically, a rotation about the *z*-axis by an angle θ would result in point P (x,y) transforming to P' (x',y') as defined below:

$$x' = x\cos(\theta) - y\sin(\theta)$$
$$y' = x\sin(\theta) - y\cos(\theta)$$

In matrix form:

$$\begin{bmatrix} x' \\ y' \end{bmatrix} = \begin{bmatrix} \cos(\theta) & -\sin(\theta) \\ \sin(\theta) & -\cos(\theta) \end{bmatrix} \cdot \begin{bmatrix} x \\ y \end{bmatrix} \text{ or } P=R.P$$

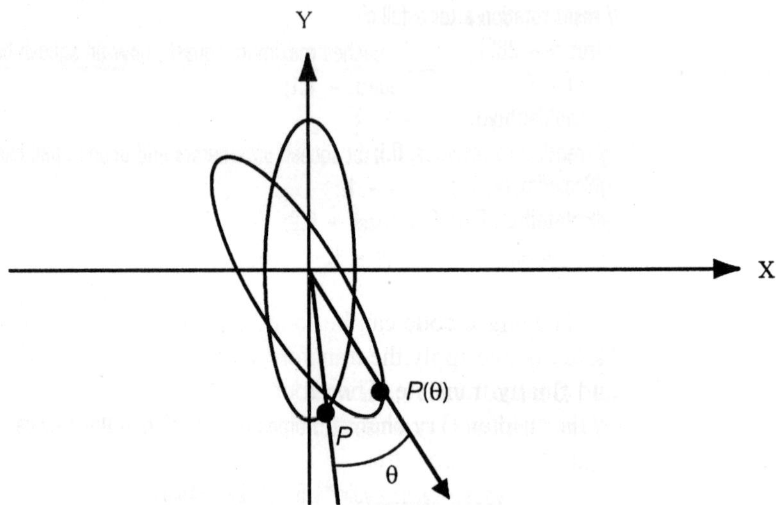

Fig.2.9: Rotation of shape

Where **R** is the rotation matrix shown above. Positive angles are measured counterclockwise from *x* toward *y*. Just like scaling, rotation occurs about the world origin. Only points at the origin are unchanged after the rotation transformation.

The OpenGL routine for rotation is

```
glRotatef(Rot, vx, vy, vz)
```

It expects the angle of rotation in degrees and a nonzero vector value about which you wish the rotation to occur. For rotations in the *xy*-plane, we would set $vx=vy=0$ and $vz=1$ (i.e., the unit vector along the *z*-axis)

Let us go back to our bouncing ball example. Of course, rotating a round ball will have no effect on the ball. Let us redefine the draw_ball routine to draw an elongated ball.

```
// Draw the ball, centered at the origin and scaled along the X axis
// It's a football!
void draw_ball() {
  glColor3f(0.6,0.3,0.);
  glScalef(1.3,1.,1.);
  MyCircle2f(0.,0.,RadiusOfBall);
}
```

To rotate the ball while it's bouncing, we add the following lines of code to our Display function:

```
rot = rot+2.;
```

```
// reset rotation after a full circle
if (rot >= 360)
   rot = 0;
glLoadIdentity();
glTranslatef(xpos,ypos, 0.);
glScalef(sx,sy, 1.);
glRotatef(rot, 0.,0.,1.);
draw_ball();
```

The entire code can be found under *Example 2_3/Example 2_3.cpp.* Note that we apply the transformations in a certain order—first rotate, then scale, and finally translate. The order of transformations does affect the final outcome of the display. Try changing the order in the example and see what happens.

Let us see how transformations are combined together in more detail.

2.3 Homogenous Coordinates and Composition of Matrix Transformations

We have seen the different vector/matrix representations for translation, scaling, and rotation. Unfortunately these all differ in their representations and cannot be combined in a consistent manner. To treat all transformations in a consistent way, the concept of *homogenous coordinates* was borrowed from geometry and applied to CG.

With homogenous coordinates, we add a third coordinate to a (2D) point. Instead of being represented by a pair of numbers (x,y), each point is now represented by a triplet (x,y,W), or in vector notation as $\begin{bmatrix} x \\ y \\ W \end{bmatrix}$

To go from homogenous coordinates back to our original non-homogenous world, we just divide the coordinates by W. So the homogenous point represented by: $\begin{bmatrix} x \\ y \\ W \end{bmatrix}$ is equal to the point $(x/W, y/W)$.

Two points of homogenous coordinates (x,y,W) and (x',y',W') are the same point if one is a multiple of the other. So, for example, (1,2,1) and (2,4,2) are the same points represented by different coordinate triples. The points with W=0 are points at infinity and will not appear in our discussions.

Because 2D points are now three element column vectors, transformation matrices used to transform a point to a new location must be of cardinality 3x3.

In this system, the translation matrix **T** is defined as

$$\begin{bmatrix} 1 & 0 & Tx \\ 0 & 1 & Ty \\ 0 & 0 & 1 \end{bmatrix}$$

Any point $P\ (x,y,W)$ that is translated by the matrix **T** results in the transformed point P' defined by the matrix-vector product of T and P:

$$P'(x',y',W') = T1.P$$

$$\begin{bmatrix} x' \\ y' \\ W' \end{bmatrix} = \begin{bmatrix} 1 & 0 & Tx \\ 0 & 1 & Ty \\ 0 & 0 & 1 \end{bmatrix} * \begin{bmatrix} x \\ y \\ W \end{bmatrix} = \begin{bmatrix} x+Tx*W \\ y+Ty*W \\ W \end{bmatrix}$$

Satisfy yourself that for W=1 this is consistent with what we learned earlier. For any other value of W, convert the coordinates to their non-homogenous form by dividing by W and verify the result as well.

The Scale transformation matrix (**S**) in homogenous coordinates is defined as:

$$\mathbf{S} = \begin{bmatrix} Sx & 0 & 0 \\ 0 & Sy & 0 \\ 0 & 0 & 1 \end{bmatrix}$$

and the rotation matrix (**R**) as:

$$\begin{bmatrix} \cos(\theta) - & \sin(\theta) & 0 \\ \sin(\theta) & \cos(\theta) & 0 \\ 0 & 0 & 1 \end{bmatrix}$$

Each of these matrices can be multiplied successively by the homogenized points of our object to yield the final transformed points. For example, suppose that points on shape P are translated by a translation matrix $\mathbf{T_1}$ and then by $\mathbf{T_2}$. The net result should be the translation $(\mathbf{T_1}+\mathbf{T_2})$. Verify that the equation:

$$P' = \mathbf{T_2}.(\mathbf{T_1}.P)$$

does indeed result in the desired transformed points.
In fact, it can be shown that this equation can be derived to be:

$$P' = \mathbf{T_2}.\mathbf{T_1}.P = (\mathbf{T_2}.\mathbf{T_1}).P = \mathbf{T_{21}}.P$$

That is, the matrix product of the two translations produces the desired transformation matrix. This matrix product is referred to as the concatenation or *composition* of the two transformations. Again, please verify for yourself that this is indeed the case. It can be proven mathematically that applying successive transformations to a vertex is equivalent to calculating the product of the transformation matrices first and then multiplying the compounded matrix to the vertex [FOLE95]. That is, we can selectively multiply the fundamental R, S and T matrices to produce desired composite transformation matrices.

The basic purpose of composing transformations is to gain efficiency. Rather than applying a series of transformations one after another, a single composed transformation is applied to the points on our object. At any given time, the current transformation state is represented by the composite matrix of the applied transformations.

Let us apply the composition principle to the ball shape under discussion. Let us go back to *Example2_2*, where we were squashing the ball upon contact with the ground. Remember that the ball seemed to jump when we squashed it. To avoid this jump, we wish to scale the ball about its bottom most point. In other words, we wish the bottommost point to stay at the same place when we squash the ball.

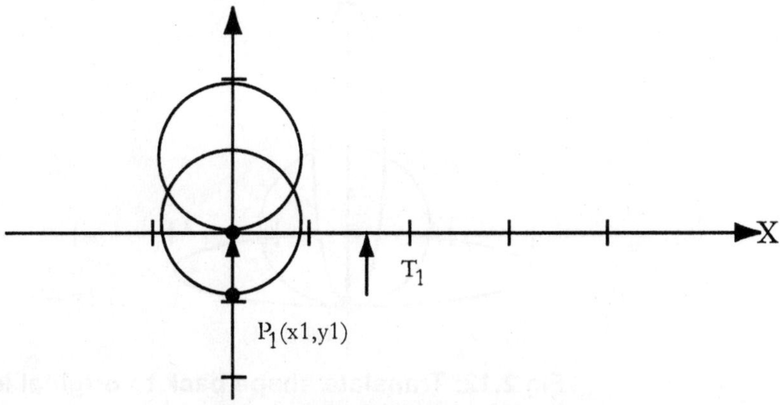

Fig.2.10: Translate shape by T1

From our study of transformations, we know how to scale the ball about the world origin (which is where the center of the ball is located). To transform the ball about its base, we can convert this problem into three basic transformations:

First, move the ball so that the desired fixed point of scaling (the base in this

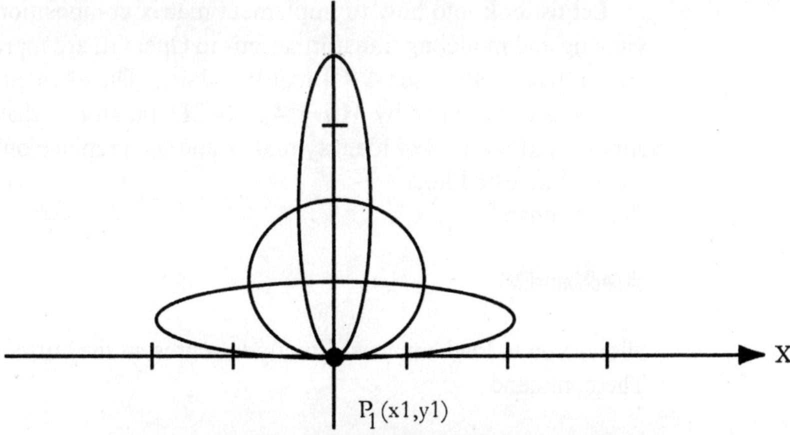

Fig.2.11: Scaling the translated shape

case) point $P_1(x1,y1)$ is at the origin, by applying translation matrix $\mathbf{T_1}$. Then we scale the shape as usual by applying a scaling transform **S**. The scaling occurs about the origin, which is now where the base of the ball is positioned.

Finally, we translate the shape back such that P_1 returns to its original point, i.e. we apply a translation transform $\mathbf{T_2} = -\mathbf{T_1}$. Viola: we have scaled the shape keeping the bottommost point P_1 (x1,y1) fixed. Effectively, we have scaled the shape about the pivot point P_1. Rotating a shape about a point works on the same principle. First translate the shape so that the point of rotation is at the origin. Apply the rotation and translate the shape back to its original location.

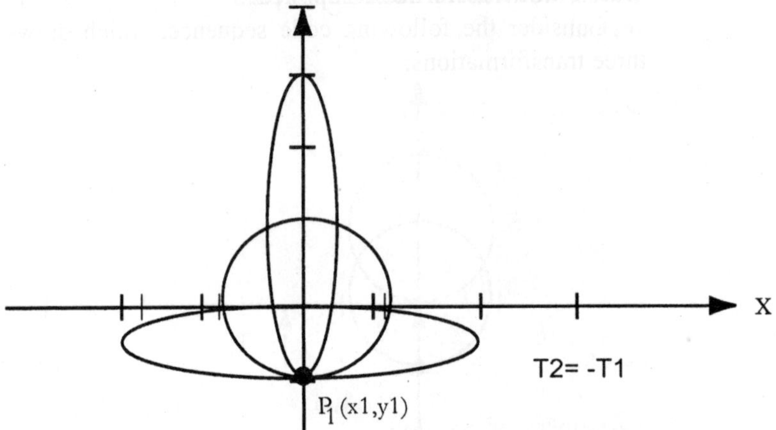

Fig.2.12: Translate shape back to original location.

The composite matrix to compute this entire transformation can be calculated as

$\mathbf{M} = \mathbf{T_2}.\mathbf{S}.\mathbf{T_1}$

This composite matrix can be applied to the shape in one step to achieve the desired result. In general, the order of composing the transforms is important. If we change the order, different results will appear.

Let us look into how to implement matrix compositions using OpenGL. All viewing and modeling transformations in OpenGL are represented as 16-element arrays (representing the 4 × 4 matrix values). The element in the *i*th row and *j*th column is represented by **M**[*i*+*j**4]. For 2D transforms that use 3x3 matrices, we simply start with a 4x4 identity matrix and then replace only those locations that we are interested in.

The command

```
glLoadMatrixf(M)
```

allows you to load a matrix directly to represent the current transformation state.

The command

```
glMultMatrixf(M)
```

multiplies the current transformation matrix by the specified matrix, **M**. Each successive glMultMatrix*() or transformation command multiplies a new 4 × 4 matrix M to the current transformation matrix **C** to yield the (new current) composite matrix **C.M**. In the end, vertices *v* are multiplied by the current matrix to yield the transformed vertex locations. This process means that the last transformation command called in your program is actually the first one applied to the vertices. For this reason, you have to specify the matrices in the reverse order of the transformation applied.

Consider the following code sequence, which draws a single point using three transformations:

```
glMatrixMode(GL_MODELVIEW);
glLoadIdentity();
glMultMatrixf(T2);          // apply transformation T2
glMultMatrixf(S);           // apply transformation S
glMultMatrixf(T1);          // apply transformation T1
glBegin(GL_POINTS);
        glVertex3f(v);              // draw transformed vertex v
glEnd();
```

Remember, we do need to set the matrix mode to specify which space the matrix operations are occurring within (in this case, the object space represented by the modelview matrix). With this code, the modelview matrix successively contains **I**, $\mathbf{T_2}$, $\mathbf{T_2.S}$, and finally $\mathbf{T_2.S.T_1}$, where **I** represents the identity matrix. The final vertex transformation applied is equal to: $(\mathbf{T_2}.(\mathbf{S}.(\mathbf{T_1}.v)))$ - notice that the transformations to vertex *v* effectively occur in the opposite order in which they were specified.

Coming back to our example, we want to scale the ball about its base. To do this, we translate the ball so that its base lies at the origin. This is done by translating the ball upward by its radius. If we initially define the transformation matrix $\mathbf{T_1}$ as the identity matrix

```
GLfloat T1[16] = {1.,0.,0.,0.,\
                  0.,1.,0.,0.,\
                  0.,0.,1.,0.,\
                  0.,0.,0.,1.}
```

then to attain the matrix to translate the ball upward by its radius, we can just set

```
T1[13] = RadiusOfBall;
Defining T1 to be {1.,0.,0.,0.,\
                   0.,1.,0.,0.,\
```

```
0.,0.,RadiusOfBall,0.,\
0.,0.,0.,1.}
```

To move the ball back down, we set

```
T1[13] - -RadiusOfBall;
```

If we define the scaling matrix **S** also to initially be the identity matrix, then to squash the ball by *sx* and *sy*, **S** would be set as

```
S[0] - sx;
S[5] - sy;
```

The transformation sequence for scaling and bouncing the ball would be as follows:

```
//Retain the current position of the ball
  T[12]- xpos;
  T[13] - ypos;
  glLoadMatrixf(T);
//Squash the ball about its base

  T1[13] - -RadiusOfBall;
  // Translate ball back to center
  glMultMatrixf(T1);
  S[0] - sx;
  S[5] - sy;
  // Scale the ball about its bottom
  glMultMatrixf(S);
  T1[13] - RadiusOfBall;
  // Translate the ball upward  so that it's bottom point is at the origin
  glMultMatrixf(T1);
draw_ball();
```

Now you will see that the ball remains on the ground when it is squashed—just as we wanted! The entire code can be found under *Example 2_4.*

An Easier Alternative

The three commands we learned earlier, namely, glTranslate, glScale, and glRotate, are all equivalent to producing an appropriate translation, rotation, or scaling matrix and then calling glMultMatrix*() with this matrix. Using the above functions, the same transform sequence for the bouncing ball would be as follows:

```
//reset transformation state
```

```
glLoadIdentity();
// retain current position
glTranslatef(xpos,ypos, 0.);

// Translate ball back to center
glTranslatef(0.,-RadiusOfBall, 0.);
// Scale ball about its bottom
glScalef(sx,sy, 1.);
// Translate the ball upward so that it's bottom is at the origin
glTranslatef(0.,RadiusOfBall, 0.);
// draw the ball
draw_ball();
```

Matrix math is very important to the understanding of graphics routines. But using matrices in actual code is involved and often tedious. It is much easier to use the glTranslate, glScale, and glRotate commands. These commands internally create the appropriate matrix and perform the necessary computations. For the rest of the book, we shall use the above-mentioned OpenGL functions in our coding examples.

Summary

In this chapter, we have covered one of the most mathematical concepts in Computer Graphics: transformations. Transformations are based on vector and matrix math. The basic transformations are of three types: translation, scale, and rotation. These can be composed together to create complex motions. Although the actual matrix math for transformations can be fairly tedious, OpenGL provides a number of functions to easily manipulate and apply transformations to objects. In the next section, we shall extend the concepts of 2D transformations learned here to the more general 3D case. This is when we will really be able to appreciate the true power of transformations.

Chapter 3

Pixels, Images and Image Files

In Chapter 1, we saw that the computer display can be treated as a two-dimensional grid of pixels. Pixels can be addressed by their (*x*, *y*) coordinates which represent their horizontal and vertical distance from the origin. All shapes (even three dimensional ones) ultimately need to identify the location of the appropriate pixels on the screen for display—a process called *rasterization*. In Chapter 2, we learned how to transform these shapes to achieve motion. We used OpenGL to render many of the concepts we learned. The two-dimensional array of colors that we created to define our image can be saved in a file for later use. This file is referred to as a raster image file.

Images need not be created only by the computer. Digital cameras, scanners, etc. can all save images to a file and send the file to the computer. The real value of saving images on the computer is what can be done to the image once it resides on the computer. You may have played with photo editors, which let you manipulate your photos-to reduce red eye for example. With good image processing tools, there is no end to the magic you can do. Photoshop by Adobe provides one of the most sophisticated image processing packages on the market. In this chapter, we will learn the following concepts:

- Image files (in particular, the BMP file format)
- How to save your OpenGL images to a file
- How to view image files within your OpenGL program
- How to manipulate pixel data from an image

3.1 Raster Image Files

A raster image file is a file that stores the mapping of pixel coordinates to color values. That is, the file saves the color at each pixel coordinate of the image. The saved image files can then be used for various purposes such as printing, editing, etc. Some commonly used image file types are: BMP, TIFF, GIF, and JPEG files.

A raster file contains graphical information about the colors of the pixels. It also contains information about the image itself, such as its width and height, format, etc. Let us look into the details of a commonly used raster file format called BMP.

The BMP File Format

The BMP file format is the most common graphics format used on the Windows platform. The motivation behind creating the BMP format was to store raster image data in a format independent of the color scheme used on any particular hardware system. The color schemes supported in BMP are monochrome, color indexed mode, and RGB mode. Support for a transparency layer (alpha) is also provided. Let us look into the structure of a BMP file. The data in BMP files is stored sequentially in a binary format and is sometimes compressed. Table 3.1 shows the basic format of the bmp file which consists of either 3 or 4 parts.

Parts of the BMP file
Header
Image Information
Color Palette (indexed mode)
Pixel data

Table 3.1: Parts of the BMP file

The first part of the file is the header. The header contains information about the type of image (BM, for BMP), its size, and the position of the actual image data with respect to the start of the file. The header is defined in a structure as shown below:

```
typedef struct tagBITMAPFILEHEADER {
        WORD   bfType;                    // must be "BM"
        DWORD  bfSize;                    // size of file
        WORD   bfReserved1;
        WORD   bfReserved2;
        DWORD  bfOffBits;              //offset to start of file
} BITMAPFILEHEADER
```

The second part of the BMP file is the image information section. Information such as the image width and height, type of compression and the number of colors is contained in the information header.

The image information data is described in the structure given below. The fields of most interest are the image width and height, the number of bits per pixel (which should be 1, 4, 8 or 24), and the compression type. Compressions are techniques applied to files to reduce their size. Compression techniques are especially useful in the case of raster image files, which tend to be huge.

```
typedef struct tagBITMAPINFOHEADER{
        DWORD   biSize;
        LONG    biWidth;
        LONG    biHeight;
        WORD    biPlanes;
        WORD    biBitCount;
        DWORD   biCompression;
        DWORD   biSizeImage;
        LONG    biXPelsPerMeter;
        LONG    biYPelsPerMeter;
        DWORD   biClrUsed;
  DWORD   biClrImportant;
} BITMAPINFOHEADER
```

The compression types supported by BMP are listed below:
0: no compression
1: 8 bit run length encoding
2: 4 bit run length encoding
3: RGB bitmap with mask

We will assume no compression of bitmap files in this book. Refer to BOVI00 if you would like more information on imaging and compression techniques.

If the image is in index color mode, then the color table information follows the information section. We will assume only RGB mode files in this book.

Last of all is the actual pixel data. The pixel data is stored by rows, left to right within each row. The rows are stored from bottom to top. The bits of each pixel are packed into bytes, and each scan line is padded with zeros to be aligned with a 32-bit boundary.

Since the bitmap image format is so simple, reading bitmap files is also very simple. The simplest data to read is 24-bit true color images. In this case the image data follows immediately after the information header. It consists of three bytes per pixel in BGR (yes, it's in reverse) order. Each byte gives the intensity for that color component-0 for no color and 255 for fully saturated. A sampling of code required to read a 24-bit image BMP file would look as follows:

```
// Read the file header and any following bitmap information...
if ((fp - fopen(filename, "rb")) -- NULL)
        return (NULL);

// Read the file header
fread(&header, sizeof(BITMAPFILEHEADER), 1, fp)
if (header.bfType !- 'MB')                          // Check for BM
reversed.
{
// Not a bitmap file
        fclose(fp);
return (NULL);
}
infosize - header.bfOffBits -
        sizeof(BITMAPFILE  HEADER);
fread(*info, 1, infosize, fp)
imgsize - (*info)->bmiHeader.biSizeImage;
// sometimes imagesize is not set in files
if (imgsize -- 0)
        imgsize - ((*info)->bmiHeader.biWidth *
                (*info)->bmiHeader.biBitCount + 7) / 8 *
                abs((*info)->bmiHeader.biHeight);
fread(pixels, 1, imgsize, fp);
```

Microsoft Windows provides the BMP file data structures definitions in <wingdi.h>. On a Linux platform, you may have to define the structs yourself. The pixels variable is a pointer to the actual bytes of color data in the image. Conceptually, you can think of it as rows and columns of cells containing the color values of each pixel. (Can you see the analogy between this structure and the frame buffer?)

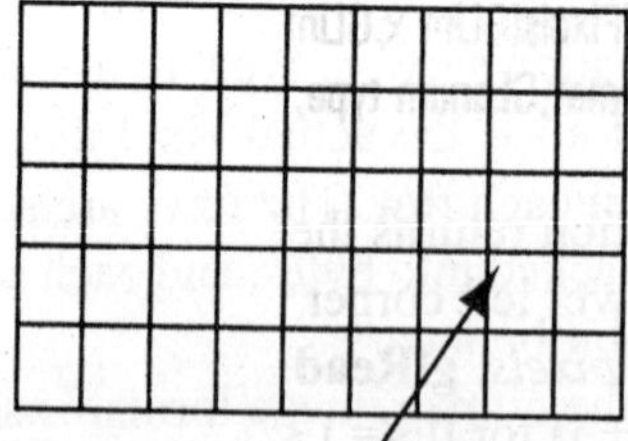

Fig.3.1: Array of RGB values

The code required to save a BMP file given the pixel data, is analogous.

We have provided C code to read and save bmp files. It can be found in the files: *bmp.cpp* and *bmp.h* under the directory where you installed the example code.. The functions have the following signatures

```
extern GLubyte * ReadBitmap(const char *filename,BITMAPINFO **info);
extern int SaveBitmap(const char *filename, BITMAPINFO *info,GLubyte    *bits);
```

where ***BITMAPINFO*** is a struct containing the BITMAPINFOHEADER struct and a color map (in the case of color index mode files).

SaveBitmap requires and ReadBitmap returns a pointer to the bytes of color information. In *bmp.h*, we have commented out the BMP header struct definitions. If you use Microsoft libraries, then you will not need to define them as they are defined in *wingdi.h*.

3.2 Bitmaps and Pixmaps

The *pixels* variable we saw in the last section is referred to as a *pixmap*. A pixmap is simply a structure that holds the color information of an image in memory. This information can then be transferred to the frame buffer for rendering on screen or can be saved back again (presumably after being manipulated in some way) as a file.

Historically, pixmap defines a color image whereas its counterpart, the *bitmap* is monochrome (black and white only). Bitmaps have only one bit for every pixel (a value of 1 is white and a value of 0 is black). But many people use the terms interchangeably. Let us see how we can use OpenGL to save our work to an image file.

Saving your work to an Image file

The OpenGL function,

glReadPixels

reads a block of pixels from the frame buffer and returns a pointer to the pixel data-yes it returns a pixmap! The exact signature for the glReadPixels function is

```
void glReadPixels(GLint x,GLint y,GLsizei width,GLsizei height,
GLenum format,GLenum type, GLvoid *pixels)
```

The function returns the *pixel* data from the frame buffer, starting with the pixel whose lower left corner is at location (x, y), into client memory pointed to be the variable: *pixels*. **glReadPixels** returns values for each pixel
(x + i, y + j) for 0 <= i < width and 0 <= j < height.

Pixels are returned in row order from the lowest to the highest row and from left to right in each row. Typically, one would like to read the entire content of the frame buffer - as defined by the extents of the viewport.

format specifies the color mode of the pixmap. GL_COLOR_INDEX for color-indexed pixels, GL_RGB for RGB pixels, and GL_BGR_EXT for RGB-mode based BMP files (since BMP files reverses the order of the R,G and B components).

type specifies the bit size of each color component. GL_BYTE or GL_UNSIGNED_BYTE is used for 8-bit values (which are used by RGB-mode images), GL_BITMAP is used for one-bit values (monochrome images) etc.

Several other parameters control the processing of the pixel data before it is placed into client memory. These parameters are set with three commands: glPixelStore, glPixelTransfer, and glPixelMap.We shall not look into the details of these calls here. Interested readers are encouraged to refer to SHRE03 for more information. For your convenience, we have defined a function:

```
GLubyte *
ReadBitmapFromScreen(BITMAPINFO **info)
```

that reads the current frame buffer, constructs an appropriate BITMAPINFO, and returns the constructed pixmap. This function is declared in the file: *bmp.h* and defined in the file: *bmp.cpp*. It is useful (but not necessary) to look at the code to understand how to setup OpenGL to read pixel values from the frame buffer.

Recall the stick figure we drew in *Example1_3. In Example3_1*, we save the stick figure to a BMP file called stick.bmp in the same directory as the executable. The code to read and save the state of the frame buffer is as follows:

```
BITMAPINFO *info;
GLubyte *pixels - ReadBitmapFromScreen(&info);
SaveBitmap("stick.bmp", info,pixels);
```

These lines of code are to be called after all drawing operations are complete, in order that the frame buffer is completely defined. The generated BMP file can now be loaded into your Microsoft Paint or any other program for further editing.

The code to draw out the stick figure and save it to a file can be found under *Example3_1/Example3_1.cpp*. You will need to compile and link it with the bmp.cpp file that we have provided.

Loading an Image file and using it in OpenGL

Let us see how we can use OpenGL to load an image file.

OpenGL provides the library function

```
glBitmap
```

to take a bitmap and transfer its contents to the frame buffer for drawing to the screen. The function

```
glDrawPixels(GLSizei width, glSizei height, Glenum format, Glenum type, Glvoid * pixels)
```

draws pixmaps onto the screen. It accepts five arguments:

Where *width* and *height* specifies the dimensions of the pixmap.

format specifies the color mode of the pixmap-GL_BGR_EXT for RGB mode BMP files.

type specifies the bit size of each color component: in our case, GL_UNSIGNED_BYTE.

And finally comes the actual *pixel* data.

The location of the bottom left corner of the pixmap on the application window is determined by the most recent call to the function

```
glRasterPos2f(x,y)
```

where *x* and *y* are the world coordinates along the *x*- and *y*-axis. This position is affected by any transformations we may apply to the current object state.

A point to note about bitmaps and pixmaps is that they are defined upside down! You can specify a negative height to invert the image.

In *Example3_2*, we read in a 64 by 64-sized pixmap. The image has a black background and a yellow colored smiley face.

Fig.3.2: A Smiley face

We keep redrawing this pixmap, bouncing it up and down in our window. Additionally, we also bounce our ball from Chapter 2 so as to demonstrate that we can still make valid OpenGL drawing requests.

```
void Display(void)
{
        glClear(GL_COLOR_BUFFER_BIT);
  glRasterPos2f(100,ypos);
        // Draw the loaded pixels
  glDrawPixels(info->bmiHeader.biWidth,
         info->bmiHeader.biHeight,
          GL_BGR_EXT, GL_UNSIGNED_BYTE, pixels);
// draw the ball at the same Y position
        draw_ball(40.,ypos);
        //dont wait, start flushing opengl calls to display buffer
        glFlush();
        glutSwapBuffers();
```

```
        // 120 is max Y value in our logical coordinate system
        ypos = ypos + 0.1;
        if (ypos >=   120.)
                ypos = 0;
        // Force a redisplay
    glutPostRedisplay();
}
```

Notice that we call the ball drawing code using its center as a parameter. This approach ensures that we draw the ball with a translated center without using the OpenGL transformation functions. The reason is that the function glRasterPos is affected by transformations. This, in turn means that it can be hard to control the exact location of our smiley face. We shall see techniques on how to handle this issue in a later chapter. The main () function reads in the desired bitmap.

```
BITMAPINFO *info;    // Bitmap information
GLubyte    *pixels;  // Actual pixel data

main(){
.

.
// read in image file as specified on the command line
        if (argc > 1)
                pixels = ReadBitmap(argv[1], &info);
}
```

The code for this example can be found in *Example3_1/ Example3_1.cpp*. Default images are located under the directory Images under the installed folder for the sample programs.

Notice the location of the pixmap versus the ball. Why is the ball located at a position lower than the image? Try to change the code so that the ball and the image bounce randomly along the *x*- and *y*-axes. For a really cool project, make the program exit when the two shapes collide. In the next chapter, we will see how to make these kinds of shapes move based on user input.

Loading more than one image

In most games of today, you see characters moving across the screen in front of a static background image. In *Example3_3*, we read in two pixmaps. One is the smiley face we just saw, and the other is a background image of the Alps. We make the background image cover the entire window by scaling it up using the function

```
glPixelZoom(x,y)
```

This function will scale up the image by the specified x and y factors.

The code to draw the background pixmap is shown below:

```
if (bgpixels) {
  glRasterPos2i(0,0);
        // scale the image appropriately and then draw the background image
        glPixelZoom(0.5,0.8);
  glDrawPixels(bginfo->bmiHeader.biWidth,
        bginfo->bmiHeader.biHeight,
        GL_BGR_EXT, GL_UNSIGNED_BYTE, bgpixels);
 }
```

The motion of the smiley face is the same as in *Example3_2*. The entire code can be found in *Example3_3/Example3_3.cpp.* In the above example, we redraw the entire background image at every call to the display function. Unfortunately, this approach leads to significantly slow performance. For complex images, the computational and redisplay rate of the computer may not be fast enough to display convincing motion. In the movie world, the images are saved and replayed at a later time, so the slow re-display rate is not an issue.

If you save each image drawn into a bitmap file and name the saved files sequentially (like *test1.bmp*, *test2.bmp*, *test3.bmp*), you can string the images together using a movie editor such as QuickTime Pro. The editor strings together the images in a sequence to generate a movie file—usually an MPEG or an MPEG-4 file. The movie file can be transferred directly to tape, or can be played by a movie player such as QuickTime. The movie file is optimized for playback and since the player is not actually calculating the images, the playback is fast enough to convey believable motion. This technique is also employed in streaming videos over the Internet.

For real-time games, where saving images and then replaying them is not an option, speed can be accomplished using other tricks. One technique is to re-draw only the pixmap of the character, leaving the background as is, using what we call overlay planes.

Graphical overlay planes are made up of additional memory positioned logically on top of the frame buffer (thus the name overlay). Typically this

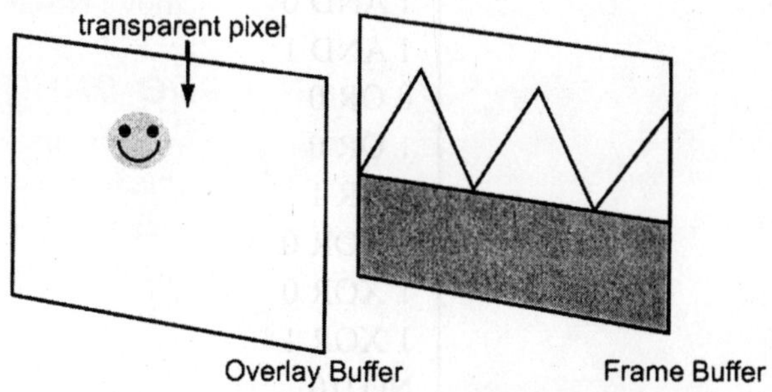

Fig.3.3: The overlay plane concept.

creates an overlay buffer that does not disturb the contents of the existing frame buffer, as shown in Fig.3.3.

Drawing and erasing in the overlay plane does not affect the frame buffer. Anything drawn in the overlay plane is always visible, and usually a color of 0 renders the overlay pixel transparent (that is, you can see the frame buffer at these values). As the frame buffer is drawn, the *x*-*y*-coordinates of each pixel are checked against the overlay buffer pixel to see if it is nontransparent. If so, then the overlay pixel is drawn instead. Popup menus are usually implemented using overlay planes.

Overlay planes are not supported natively by most standard graphics cards, so we do not implement them in this book. Refer to [SHRE03] for more details on how to develop code using overlay planes.

There are other ways we can get around the speed issue when performing intensive pixel copying and erasing. A common technique is to use logical operations on the pixels. Logical operations form the basis for many image processing techniques, so we will devote the next section on it.

3.3 Computer Display Systems

Logical operations are performed between two data bits (except for the NOT operation, which is performed on one). Bits can be either 1 or 0 (sometimes referred to as TRUE and FALSE). The most basic logical operators that can be performed on these bits are: AND, OR, XOR, and INVERT/NOT. These operators are essential to performing digital math operations. Table 2 shows the values (truth table) for these operations.

Since pixels are nothing more than a series of bits in a buffer, they can be combined using bitwise logical operations (that is, we apply the logical operations to each corresponding bit). The operations are applied between the incoming pixel values (source) and those currently into the frame buffer

Operation	Value
0 AND 0	0
1 AND 0	0
1 AND 1	1
0 OR 0	0
1 OR 0	1
1 OR 1	1
0 XOR 0	0
1 XOR 0	1
1 XOR 1	0
NOT(0)	1
NOT(1)	0

Table 3.2: Logical operations

(destination). The resultant pixel values are saved back in the frame buffer. For true color pixels, the operations are performed on the corresponding bits of the two pixel values, yielding the final pixel value.

Logical operations are especially useful on bit-blt type machines. These machines allow you to perform an arbitrary logical operation on the incoming data and the data already present, ultimately replacing the existing data with the results of the operation, all in hardware. Since this process can be implemented fairly cheaply and quickly in hardware, many such machines are available. All gaming hardware, such as Nintendo and Xbox, support these operations in hardware for fast implementations of pixel/bit copying and drawing.

In Fig. 3.4, we show the OR operator applied to two monochrome (single bit) pixel values. The bits with value 1 are shown in a white color, and those with value 0 are shown in black.

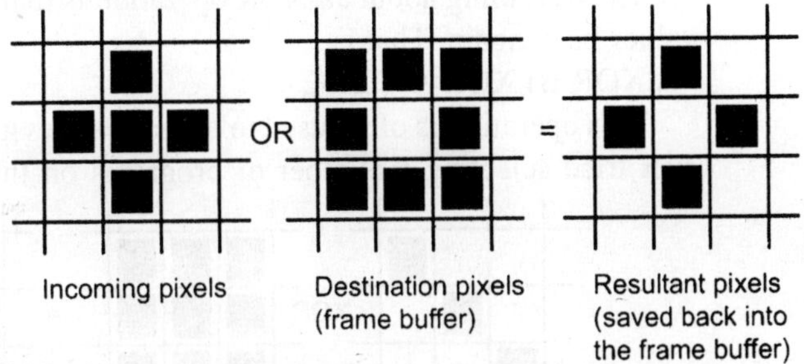

Fig.3.4: The OR operation

ORing the source over the destination combines the two pixel values. Zero-value pixels in the source are effectively transparent pixels, since the destination retains its pixel value at these points. The same operation can be performed on RGB pixels by doing a bitwise OR operation.

An AND operator uses the source as a mask to select pixels in the destination and clears out the rest, as shown in Figure .

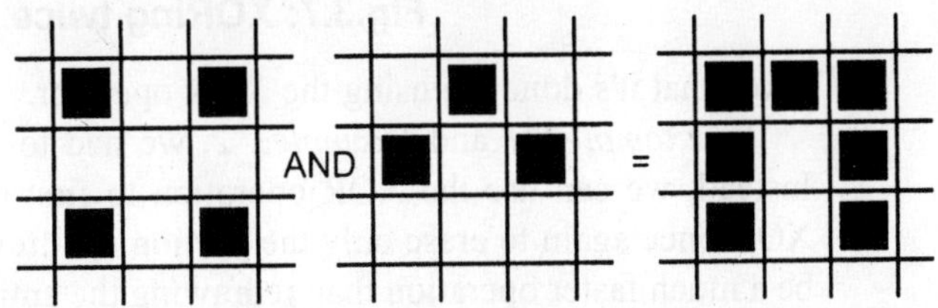

Fig.3.5: The AND operation

Image compositing, where two images are combined in some manner to create a third resultant image, make use of these two operations heavily.

XORing the source over the destination guarantees that the source stands out. This technique is used very often in rubber-banding, where a line or a rectangle is dragged around by the user on top of a background image.

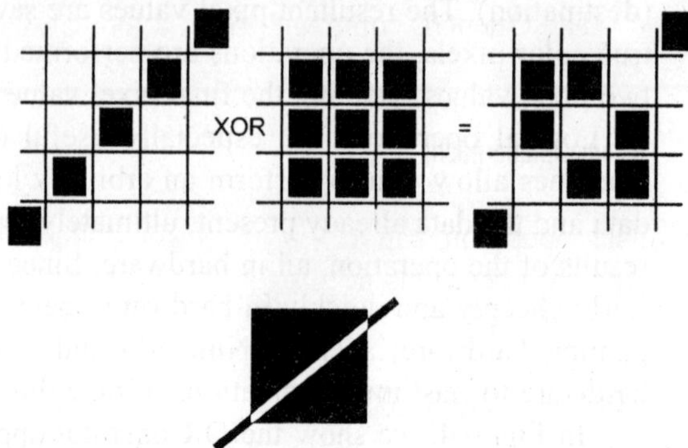

Fig.3.6: Rubber-banding: XOR of a line with the background image

The neat thing about an XOR operation is that two XORs generates the same values back again. That is,
(A XOR B) XOR A = B.

This operation is often used to to erase a previously drawn image. If you have ever tried selecting a number of programs on the Windows desktop, now you

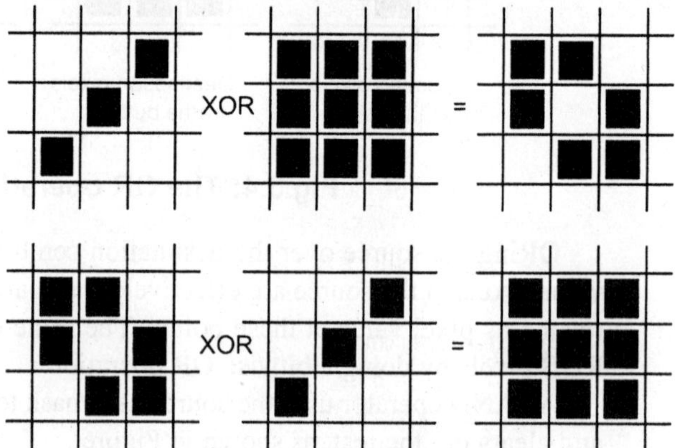

Fig.3.7: XORing twice

know that it's done are using the XOR operator.

In *Example3_1* and *Example3_2*, we had to redraw the entire background. Instead, we can use the XOR operation to first draw the smiley face and then XOR once again to erase only the portion we drew into. This technique tends to be a much faster operation than redrawing the entire background and is therefore used extensively in gaming.

A note of caution!. This approach works best for constant color images, since the colors do get flipped on the first XOR. For multicolor images, the XOR operation may result in psychedelic colors being produced. As a result, XOR should be used only when the image colors are designed to work for it. Let us see how we can modify *Example3_1* to use logical operations.

The OpenGL command to set a logical operation to be applied to incoming values is

```
glLogicOp(operation)
```

The operation can be of type GL_XOR, GL_AND, GL_OR or GL_NOT. Logical operations need to be enabled before they will have effect. To do this, call the function

```
glEnable(GL_COLOR_LOGIC_OP)
```

glDisable() will disable logical operations from occuring.

In the display code shown below, we clear out the background only once at startup. Then we use two XOR operations, one to draw the smiley face and then another to erase it. When you run the program you will notice that this process does flip the color of the smiley face!

```
if (FIRSTDISPLAY) {
// Only clear the background the first time
        glClear(GL_COLOR_BUFFER_BIT);
FIRSTDISPLAY = FALSE;
        // enable logical operations
glEnable(GL_COLOR_LOGIC_OP);
        // Set logical operation to XOR
glLogicOp(GL_XOR);
} else {
// XOR incoming values with pixels in frame buffer
        // the next two lines will  erase the previous image drawn
glRasterPos2f(prevx,prevy);
glDrawPixels(info->bmiHeader.biWidth,
 info->bmiHeader.biHeight,
  GL_BGR_EXT, GL_UNSIGNED_BYTE, pixels);
}
// the next two lines draws in a new (XORed) image
glRasterPos2f(xpos,ypos);
glDrawPixels(info->bmiHeader.biWidth,
 info->bmiHeader.biHeight,
  GL_BGR_EXT, GL_UNSIGNED_BYTE, pixels);
```

An additional bonus: since the smiley face has a black background (recall that black is the transparent color by default), the resultant display only draws the face, not the entire rectangular image. The entire code can be found in *Example3_4/Example 3_4.cpp.* Try using the background image of the Alps and see whether the XOR operation results in a desirable output.

3.4 Image Enhancements

The real value of saving images on the computer is image processing - what can be done to the image once it is resident on the computer. Almost all production studios use a processing tool to perform final image touchups, called post-production, to enhance the quality of their images. These images can be from live action footage or from computer-generated imagery. Effects like blending, compositing, and cleaning up pixel values are used routinely in productions. The idea behind image processing is simple: manipulate the pixel information of the given image using some mathematical functions and save the result in a new image file. We discuss two techniques in this section, namely, compositing and red-eye removal.

Refer to [PORT84] for information on other image processing techniques.

Image Compositing

Compositing is the most widely used operation in the post-production of films, commercials, and even TV broadcasts. The basic operation in image compositing is to overlay one (usually nonrectangular) image on top of the other. Recall from *Example3_2*, when we drew the smiley face on top of the background, we displayed the entire image, not just the face, as we would have liked. XORing flips the colors, so that is not always a good solution either. Compositing to the rescue!

Various techniques are used for compositing images. One popular technique is the blue screen process. First, the subject is photographed in front of an evenly lit, bright, pure blue background. Then the compositing process, whether photographic or electronic, replaces all the blue in the picture with the background image, known as the background plate. In reality, any color can be used for the background, but blue has been favored since it is the complementary color to flesh tone. The source and destination images can come from live action images or be computer generated. Jurassic Park employed compositing extensively to composite CG-generated dinosaurs in front of real live shots.

The essential compositing algorithm is as shown below. It designates Blue or the desired color as transparent. The source image is copied on top of a defined destination image as follows:

```
for y - 1 to height
for x - 1 to width
if image[x, y] <> transparent pixel then
            copy image[x, y]
else
            leave the destination unchanged
```

Fig. 3.8: Compositing Algorithm

The only problem, of course, is that "transparent" color can't be used in the source image. Another technique for compositing, being used more frequently now makes use of a mask or stencil. This technique uses logical operations as follows:

- Create a mask of the image. Many production houses create the mask as part of the alpha channel of the image. The mask essentially identifies the desired areas of the source image.
- AND the source with the mask (not always required).
- AND the background with NOT of mask
- OR the results to produce the final composite.

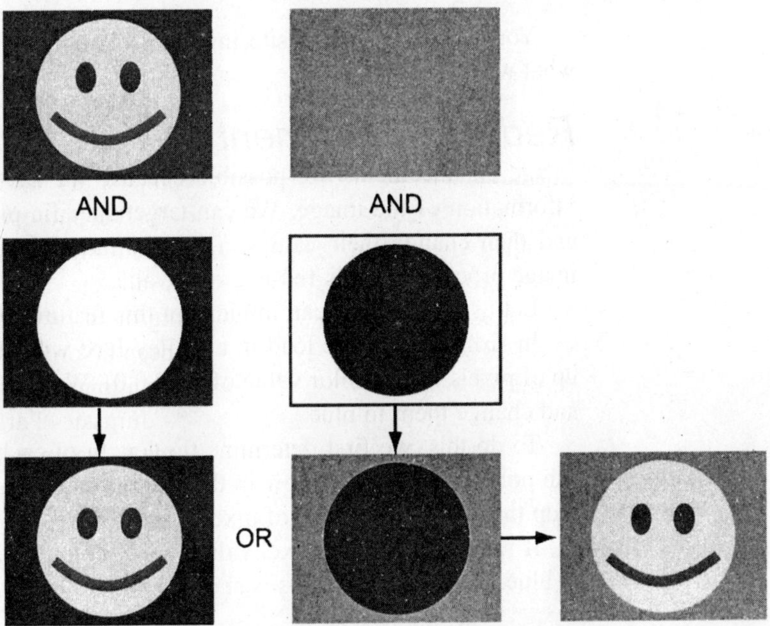

Fig. 3.9: Image Composition using masks

We show below, sample code for compositing the smiley face on top of the background image of the Alps. The mask was created as a BMP image using Adobe Photoshop. The entire code can be found under *Example3_5/Example 3_5.cpp.*

```
    // First copy background image into frame buffer
glLogicOp(GL_COPY);
glRasterPos2i(0,0);
glPixelZoom(0.5,0.8);
glDrawPixels(bginfo->bmiHeader.biWidth, bginfo->bmiHeader.biHeight,
  GL_BGR_EXT, GL_UNSIGNED_BYTE, bgpixels);

// perform an AND with destination and NOT(source - mask)
glLogicOp(GL_AND_INVERTED);
glRasterPos2f(xpos,ypos);
glPixelZoom(1.0,1.0);
glDrawPixels(maskinfo->bmiHeader.biWidth, maskinfo->bmiHeader.biHeight,
  GL_BGR_EXT, GL_UNSIGNED_BYTE, maskpixels);

// Perform an OR with source- smiley
glLogicOp(GL_OR);
glDrawPixels(info->bmiHeader.biWidth, info->bmiHeader.biHeight,
  GL_BGR_EXT, GL_UNSIGNED_BYTE, pixels);
```

You will see a composite image of the smiley face on top of the Alps-just what we wanted!

Red-Eye Treatment

Image enhancements are possible because we have access to the actual pixel information of the image. We can target specific pixels based on some criteria and then change their values. You may have used a very common feature in image processing tools-red-eye removal.

Let us see how we can implement this feature for ourselves.

In *Example3_6*, we load in a smiley face with red eyes. The eyes are made up of pixels with a color value of (255,0,0). We will target pixels with this color and change them to blue.

To do this, we first determine the length of each row. Using the length, we can point to any desired row in the pixmap as the pixel_row variable. Then we loop through the columns of pixel values in this row.

If we determine the pixel value in any column has a red color, we change it to blue. The code to do this is shown below:

```
// length of each row -
//remember that the color component values are padded to a 32 bit value
  int length - (info->bmiHeader.biWidth * 3 + 3) & ~3;
  // for each row
  for (row-0; row < info->bmiHeader.biHeight; row++) {
    pixel_row - pixels + row*length;
          //Now we can loop through all the pixel values in this row as shown
```

```
for (col=0; col< info->bmiHeader.biWidth; col++, pixel_row+=3) {
                    // Is the pixel value at this row and col red?
                    // Remember!! BMP files save color info in reverse order (B,G,R)
                    if (pixel_row[0] == 0 && pixel_row[1] == 0 && pixel_row[2] == 255){
                    // yes, change the pixel color to  blue.
                              pixel_row[0] = 255; pixel_row[2] = 0;
                    }
          }
}
```

The entire code can be found in *Example3_6/Example3_6.cpp*. When you run this code, the display will show a blue-eyed smiley face! You can save the result to a file to verify that we did indeed perform red-eye removal. This is an oversimplification of the actual mechanics behind red eye reduction, but you get the idea. Now when you use a processing tool to get rid of the reds, you will know what kind of technology is working in the background.

Summary

In this chapter, we have seen how images are stored on the computer. Any kind of graphics, whether a visually realistic simulation of the environment or a simple 2D drawing, is eventually saved as a 2D raster image on the computer. Storing and viewing these images is just the tip of the iceberg: we can use these images in our own projects and even modify them to achieve desired results and special effects. The techniques we have discussed here are used extensively in production suites for image processing and gaming. They form the basis for the more complicated image enhancements and effects used in the industry. In the next chapter, we shall put together all our current working knowledge to design and develop a 2D game.

Chapter 4

Let The Games Begin

In the last few chapters, we learned the basics of 2D computer graphics. We saw how to draw simple shapes using OpenGL and how to assign colors to them. We learned how to make these shapes move about in our CG world. We also saw how to save, load, and modify computer images. Let us make things interesting by using all our knowledge to design a 2D game.

When it comes to games, we need real-time graphics—that is, the program has to be interactive and have an immediate response to input. You press a button, and the graphics changes. You drag the mouse, and a character moves in response. In this chapter, we shall learn some techniques to interact with user input in real time.

We begin by first exploring what a computer game really is and discussing the traditional processes used in designing one. Our discussions on game design will pave the way for a real game implementation.

In this chapter, you will learn the following:

- What a game is
- What steps production houses follow when producing a game
- How to follow this process to design and implement your own game

4.1 What is a Game?

If we desire to design a game, we must define what is meant by the word *game*. A game, like a story, can be thought of as a representation of a fantasy that the reader/player is invited to experience. A story is a vehicle for representing fantasy through the cause-and-effect relationships suggested by the sequence of facts it details. Games attempt to represent fantasy by a branching tree of sequences. It allows the player to create his or her own story by making choices at each branch point.

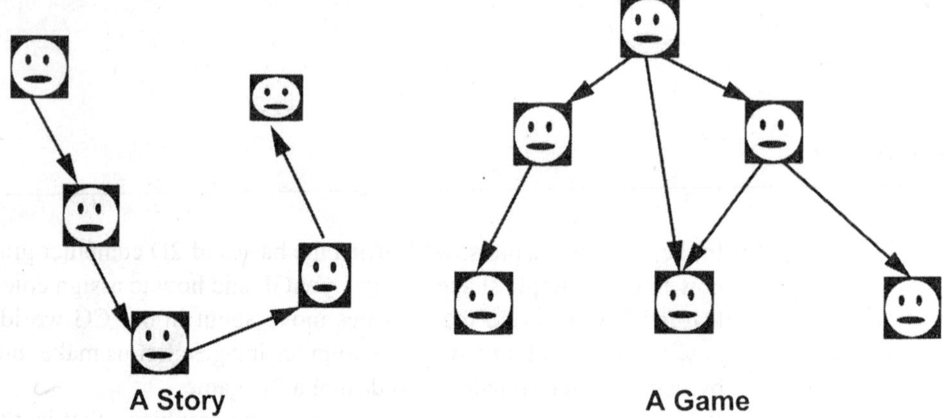

Fig.4.1: A Story vs. a Game

Game-playing requires a player. The game designer works to produce a game so the player is enticed to play it. Hence, the human player is the primary concern of the game designer. Why do people play games? What motivates them? What makes games fun? The answers to these questions are crucial to good game design. Although games are played using different media—board games, card games, etc.—we are currently interested only in games played on the computer.

There are many motivations for humans to play games: learning, fantasy/exploration, social acceptance, need for acknowledgment, etc. Some factors motivate a person to play games; other factors help that person select a particular game. A game cannot be fun if its factors do not satisfy the motivations of the player. So when designing a game, it is important to understand the motivations of the target audience. Different games motivate different kinds of audience. An intellectual audience may be motivated to play a game for mental exercise, so they would probably prefer those games that offer a greater mental challenge than do other games. If the target audience is motivated by action and combat, then that is what the designer should integrate into the game.

Sensory gratification is usually the deciding factor between games with the same intent. Good graphics, color, animation, and sound are all valued by game players. These elements support the fantasy of the game by providing sensory "proof" of the game's reality.

Keeping the above in mind, let us work through designing and implementing a game for ourselves.

4.2 Game design

Let us work through the design of a game following the standard steps used in professional game design. These steps force the designer to think through the important issues about the game before beginning implementation of the game. This is a *very* important process, since if the goals of the game are not clearly identified, the game will probably fail to meet its primary objective of motivating players.

1. **Audience**

The first step in our game design process is to identify the target audience. For the purposes of this exercise, let us pick children as our audience. In particular—boys between the ages of 8 and 12.

Once the audience is chosen, research is carried out to identify the likes and dislikes of this group. Say we find out that among the many things the boys love to do, aiming and shooting at objects rank high.

Let us define our game to be a shooting game. The goal of the game will be to provide the player with the opportunity to aim and fire at objects. The game will also enhance the player's fine motor skills and hand-eye coordination.

2. **Environment**

Next, we must determine the environment in which the game will be played. We have already defined the game to be a shooting game, but what characters are playing in the game and what environment are they set in?

Let us define the game to be set in outer space. We need a character that will be controlled by the player. This character will allow the player to shoot. We also need a character to shoot down. We further want these characters to blend into the environment selected.

With this in mind, we define a user-controlled spaceship as the shooter, and meteors that appear randomly as the target. In order for the game to have a good story, we probably need a third character which is the object that the user is trying to protect. A silhouette of planet Earth would be a good choice, since everyone is motivated to protect it from being hit by meteors!

3. **Interactive screenplay**

The screenplay, if appropriate, contains the dialogs and the storyline implemented in the game.

For this game, let us define a simple screenplay. The game opens with a space setting simulated with stars and sky. The tip of planet Earth is at the bottom of the scene. A spaceship is hovering on top of the earth. The player can move the ship left and right and shoot bullets upwards. Meteors, generated by the game, randomly appear in the sky. The player needs to shoot down these falling meteors. If a bullet hits a meteor, it is destroyed. If the meteor hits Earth, then the game ends with an explosion.

The interactive logic of this game can be shown as follows:

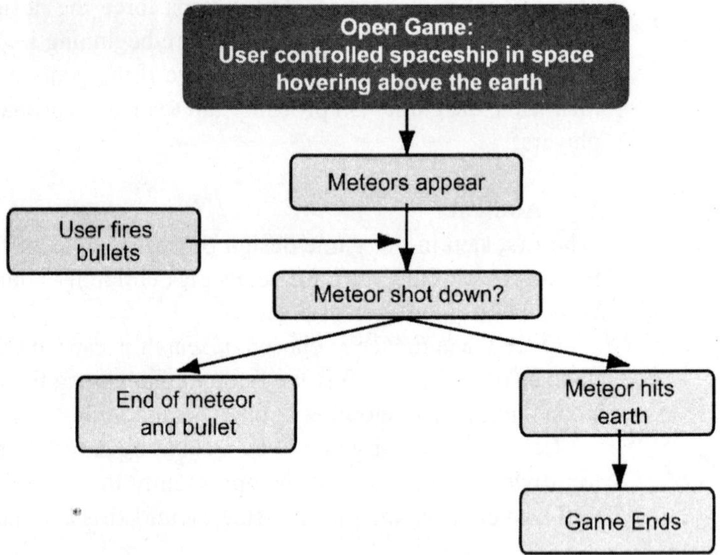

4. **Storyboard**

Storyboards aid in the design process by illustrating the game in motion. Usually, storyboards are quick sketches of key moments during the lifetime of the game. Any quirks or inconsistencies in the game show up at this step.

Storyboard illustrate only the main points of the story. For this game, the following key points would be illustrated (the actual illustrations are left as an exercise to the reader):

- The opening scene: The spaceship hovering:
- Meteors appearing randomly in space
- User positioning spaceship and firing at meteor
- End of game when meteor hits earth

5. **Implementation**

Finally, implementation issues detail the choices made to develop the game. Some of the choices are as follows:

- *Is the game 2D or 3D?*

We have no choice for the moment: it will be a 2D game.

- *What kind of look and feel does the game have?*

The game is set up to be in space. We are going for a cartoonish look as opposed to abstract or surrealistic.

- *How are the characters implemented?*

The sky and stars can be implemented using a background image. We can probably get away with defining the earth to be part of this background image, as long as we know its coordinate data in order to determine collisions.

Fig.4.2: The Background

The spaceship will also be rendered using an image file as shown.

Fig.4.3: The Spaceship

The meteors are merely *point* shapes that randomly appear in the sky and fall down towards Earth. Bullets are also point shapes that shoot upwards when the user clicks the right mouse button.

- *What are the input methods for the user?*

In arcade games, a joystick enables the user to control the game. A joystick works very similarly to a mouse. It allows the user to move characters left, right, up, and down, and allows the user to press a button to fire. For our game, the mouse will provide the input parameters. The user drags the mouse with the left button down in order to move the spaceship. Clicking on the right mouse button controls firing the bullets.

- *What are the tools needed to develop the game?*

C++ and OpenGL We will develop the game in C++ since this language lends itself very nicely to game development.

- *What platform will the game run on?*

Windows (but you can compile and run it on any platform)

4.3 Implementing the Game

By now you are probably more than ready to roll up your sleeves and start implementing the game. Let us begin. The algorithm for the game (based on the interactive logic chart we drew up earlier) is as follows:

```
till game ends do {
        monitor user input to update spaceship location
        monitor user input to generate bullet
        randomly create meteors
        for all meteors {
                update meteor location
                if meteor hits earth {
                        Display explosion
                        End Game;
                }
                if meteor has gone out of screen
                        destroy meteor
                else
                        draw meteor
        for all bullets {
                        update bullet location
                        if bullet hits meteor{
                        destroy meteor
                        destroy bullet
                } else if bullet is out of screen
                                destroy bullet
                else
                                draw bullet
                }
        }
        drawspaceship
}
```

Let us begin developing the game by first defining the characters that play in our game.

The Characters

Object oriented programming lends itself very nicely to game programming, since all the functionality of the characters can be encapsulated within a class. We shall use simple C++ classes to create the characters in the game. If you have never used C++ before, do not despair! The code is very similar to C and you will find it very easy to follow the logic of the code. Just think of a C++ class as a C struct.

The Meteor object

First, let us consider the meteor object. A meteor has a location, which identifies it's (x,y) coordinates on the screen. Based on the location, the meteor object needs to be able to draw itself. It needs to be able to update its location within every loop of the code—sometimes referred to as a *tick* of the game. It also needs to be able to identify whether it is out of the screen (in which case we destroy it) or whether it has hit planet Earth.

The meteor class can defined as follows

```
class Meteor
{
public:
  Meteor(int uID);
  virtual ~Meteor();
  void Draw();                    //Draw  the meteor
  void DrawCollision();           //Draw the meteor that has collided
  void Update();                  //Update the meteor location, for this example simply
                                  decrement it's y-coordinate in order to move it down
  bool HitEarth();                //returns true if the meteor has hit Earth
  bool OutOfScreen();             //returns true if the meteor is out of screen
  GLint ID;                       // unique ID to identify the meteor
GLfloat* GetLocation() {return location;}
private:
  GLfloat location[2];            // meteor's location. location[0]=x-, location[1] =y- coordinate
};
```

The location array stores the current (x,y)-coordinates of the meteor. For this example, we let all meteors fall straight down along the y-axis.

The update function merely decrements the location[1] variable in order to make the meteor fall down. You can experiment with giving meteors an x-direction as well.

The *Draw* routine draws a white colored point (glVertex) at the meteor's current location, whereas the *DrawCollision* routine draws a bigger red colored point to indicate a collision.

The functions *HitEarth* and *OutOfScreen* test the meteor's current location to determine if it has hit Earth or is out of the boundary of the defined world coorindates. It returns a TRUE value if the test is positive.

We use the ID variable to uniquely identify each meteor.

The entire code for this object can be found under the directory *Example4_1*, in files *Meteor.h and Meteor.cpp.*

The function *createMeteor* is used to randomly create meteors during the game.The function generates a random number between 0 and 1 and only creates a meteor if this number is less than a constant. This is to ensure we do not have too many meteors being generated.

```
void createMeteor() {
 if ( ((float)rand()/RAND_MAX) < 0.99)
        return;
 SYSTEMTIME systime;
 GetSystemTime(&systime);
 int ID = systime.wMinute*60+systime.wSecond;
  Meteors[ID] = new Meteor(ID);            //create a meteor with identifier=ID
}
```

Meteors are created only if a random number generated between 0 and 1 is less than a constant value (0.99). Decreasing this constant will cause more meteors to be created. Each meteor is instantiated with a unique ID that is an integer value. The ID of the meteor is derived from the minute and second of the current time and is unique just as long as the player does not play the game for more than an hour. This ID is used as a key to insert and locate the meteor in a meteormap:

```
typedef map<int, Meteor*> MeteorMap;
static MeteorMap Meteors;
```

We use an STL container—a map in this case—for storing the meteors. STL maps are an easy way to store objects in an array indexed by a key. In this case, the ID is the key used to store and retrieve the corresponding meteor object from the MeteorMap. STL iterators are used to loop through STL containers. More information on how to use STL can be found in **GLAS95**. You can easily change the code to manage your meteor container as a linked list.

The Bullet Object

The class encapsulating the bullet object is defined similarly to the Meteor class we just saw. The Bullet class has an additional function to test for collisions between the bullet and a given meteor.

```
class Bullet
{
public:
        Bullet(int uID, int x, int y);      //Define a new bullet at location(x,y).
```

```
virtual ~Bullet();
  void Draw();                      //Draw Bullet
  void Update();                    //Update Bullet location
  bool OutOfScreen();               //returns true if Bullet is out of screen
  bool Collide(Meteor *);           //returns true if Bullet collides with Meteor
  GLfloat ID;
private:

  GLfloat location[2];              //(x,y) location of bullet
};
```

The Draw routine draws a red-colored point (glVertex) at the bullet's current location, whereas the DrawCollision routine draws a bigger red colored point to indicate a collision. The update function merely increments the location[1] variable, thereby moving the bullet up.

All bullets created are entered into a BulletMap and located by their ID. The code for this class can be found under the folder *Example4_1* in files *Bullet.cpp* and *Bullet.h.*

```
typedef map<int, Bullet*> BulletMap;
static BulletMap Bullets;
```

Bullets are created when the user clicks the right mouse button. We shall see shortly, how to monitor for mouse events. The createBullet function is called when the right mouse button is clicked. The function is defined as follows:

```
// Create a bullet and add it to a Bulletmap
void createBullet(int x, int y) {
  SYSTEMTIME systime;
  GetSystemTime(&systime);
  int ID = systime.wMinute*60+systime.wSecond;
  Bullets[ID]= new Bullet(ID,x,y);
}
```

The Spaceship

Last of all let us consider the spaceship object. Like the meteor and bullet, the spaceship has a location variable to determine its (x,y)-coordinates.

```
class SpaceShip
{
public:
    SpaceShip();
    virtual ~SpaceShip();
    void SetPixels(BITMAPINFO *ssinfo, GLubyte   *sspixels); // set the pixels and bitmapinfo
```

```
                                            for spaceship
    void Draw();                            //Draw the spaceship
    void Update(GLfloat x, GLfloat y);      //update location of spaceship
    GLfloat GetXLocation(){ return location[0];}
    GLfloat GetYLocation() {return location[1];}
    void StartMove();                       //start moving the spaceship
    void StopMove();                        //stop moving the spaceship· update of
                                            spaceship will not do anything
  private:

    GLfloat location[2];

    GLubyte *mypixels;                      //pixel info of spaceship
    BITMAPINFO *myimginfo;                  //bitmap info of spacehsip image

    bool b_updatess;                        //are we updating location of spaceship

  };
```

The function SetPixels is used to initialize the pixmap of the spaceship. The Draw routine of the spaceship object draws a pixmap at the object's current location.

The spaceship is in motion when the user clicks down the right mouse and drags the mouse around. The moment the user releases the button, the spaceship stops moving. The Boolean variabe b_updates keeps track of whether the spaceship is in motion (and hence whether it's location needs be updated).

The Update function is called only when b_updates has a value of TRUE. It takes as parameter the current Mouse location, and draws the spaceship at this location. For this exercise, we move the spaceship only left and right, so only the *x*-coordinate is required for motion.

User Input

OpenGL and the GLUT library provide a number of functions to create event-handling threads. An event-handling thread is a separate thread of execution that listens for events such as user input, timers etc. If an event is detected, a call is made to the appropriate "callback" function. The callback function can then handle the event as desired. We saw event handlers for the Display and Reshape events in the previous chapters. Event-handling threads are the crux of all gaming programs. The GLUT function used to create a mouse event-handling thread is

```
glutMouseFunc(void(*func)(int button, int state, int x, int y))
```

glutMouseFunc sets the mouse callback for the current window to be the function *func*. When a user presses and releases mouse buttons in the window, each press and release generates a call to this function. The callback function receives information about which button was clicked (GLUT_RIGHT_BUTTON or GLUT_LEFT_BUTTON), the state of the button (GLUT_UP or GLUT_DOWN), and the *x,y* location of the mouse click within the window (in the world coordinate system). In our example, we define our mouse handling callback function to be

```
void getMouse(int button, int state, int x, int y)
```

When the right mouse button is clicked up, we want to fire a bullet. The bullet starts off at the center of the spaceship. If the spaceship was being dragged around, we also want to stop moving it. If the button is clicked down, we want to start moving the spaceship.

```
void getMouse(int button, int state, int x, int y){
 // If right mouse is clicked up, create a bullet located at the center of the spaceship image. Stop spaceship from moving
        if (button == GLUT_RIGHT_BUTTON && state == GLUT_UP) {
                createBullet(spaceship->GetXLocation()+25,spaceship->GetYLocation()+25);
                spaceship->StopMove();}
        else {
                // Start spaceship motion
                spaceship->StartMove();
        }
}
```

The command

```
glutMotionFunc (void (*func)( int x, int y))
```

is used to set the motion callback function. The motion callback for a window is called when the mouse moves within the window while one or more mouse buttons are pressed. We want the location of the spaceship to be updated when the user drags the mouse around. We define our motion callback function as follows

```
void getMouseMotion(int x, int y){
        // update spaceship when mouse is moved.
        // spaceship motion is enabled only when button is clicked down
        spaceship->Update(x, SCREENMAXY-y );
}
```

Notice something odd? We don't use the same *y*-coordinate that we receive from the mouse click! Our world coordinates set the origin to be at the bottom leftmost corner of the window. *Y* is increasing as we go up the screen. However, Windows uses (0,0) at the top left corner of the window so *y*- is increasing going down! The location of mouse events is in the coordinate system of the window and needs to be reversed to map correctly into our world! The code for this class can be found under *Example4_1*, in files *SpaceShip.cpp* and *SpaceShip.h*.

Timer Callbacks

Similar to mouse event callbacks, GLUT has the ability to register timer callbacks. Timer callbacks get triggered at regular intervals of time—the "ticks" of the game. This mechanism is extremely useful in games, when you want the game logic to be called at regular ticks.

The GLUT function

```
void glutTimerFunc ( unsigned int msecs , void (*func)(int value), value);
```

registers a timer callback to be triggered in a specified number of milliseconds. It expects a pointer to the event-handling function as a parameter, as well as any integer value to be passed to this function. In the init() code of our program, we register a timer callback as

```
glutTimerFunc(100, timer, 0);
```

This means that the function *timer* (listed below) will be called after 100 millseconds. The timer function creates meteors randomly, forces the Display function to be called and finally makes another call to re-register the timer callback, starting off the process all over again.

```
void timer(int value)
{
 createMeteor();
          // Force a redisplay, Display function contains main game logic
 glutPostRedisplay();
          // Restart the timer
 glutTimerFunc(10, timer, 0);
}
```

Putting It All Together

Now that we have all the pieces defined, let us look into the guts of the main code. The main Display() function follows the logic of the algorithm listed previously. The exact C++ implementation can be found in *Example4_1/Example4_1.cpp.* If a meteor *m* hits Earth, we call the function

```
void EndProgram(Meteor *m)
```

This function first displays a colliding meteor (a dot displayed in a red color), then loads in a series of images to depict the Earth exploding, and finally exits from the program. The initialization calls that set up all our event handlers, seed the random number generator, and create the spaceship are defined in the function init:

```
void init(void){

 // Timer callback
 glutTimerFunc(100, timer, 0);
 // Define mouse callback function
 glutMouseFunc(getMouse);
 // Define Mouse Motion callback function
 glutMotionFunc(getMouseMotion);

 // random number generator seeded by current time
 SYSTEMTIME systime;
 GetSystemTime(&systime);
 // Seed random number generator
 srand(systime.wMinute*60+systime.wSecond);

 //Create the spaceship
 createSpaceShip();

}
```

All the classes and functions can be found under the directory, *Example4_1*. All the images can be found under the directory, Images/Chapter4.

The program is defined in the simplest manner possible for easy readability. It is not the most efficient code. There are a number of tricks you can do to speed up the program: Redrawing using XOR (remember, care must be taken to make sure color patterns work during XOR), minimizing the number of loops, etc.

You can also add many more bells and whistles to get cool effects, such as an animation of the planet exploding instead of a display of a few static images, a fancier meteor object, etc. We leave this part of the game as an exercise for the reader.

Summary

In this chapter, we have combined all of our working knowledge in Computer Graphics to develop a computer game. We have learned the fundamentals of game design and the processes involved in designing and developing a game. In chapter 11, we will have the opportunity to design a 3D game.

Section 2

It's 3D Time!

We hope you have enj
understanding of the 2D
into the world of 3D grap

The mathematics in a
Most concepts in 3D are
Positions of objects are t
Transformation equation

But 3D graphics is a
begins our quest into vis
computer and then comp

3D Productions are
rendering and animation.
the basis for this section.

In Chapters 5 and 6, v
in Chapter 7, we will int
the field. Chapter 8 is en
in the graphics industry.

By the end of the sec
it on the computer to pr
ready for the 3D roller c

Section 2

It's 3D Time!

We hope you have enjoyed your ride through the 2D world. With a solid understanding of the 2D math involved in this world, we are ready to venture into the world of 3D graphics.

The mathematics in a 3D world is an elegant extension of its 2D counterparts. Most concepts in 3D are developed from 2D by simply adding a third (z-) axis. Positions of objects are now described as (x,y,z) triplets instead of just (x,y) pairs. Transformation equations are defined against three axes, not two.

But 3D graphics is a lot more than just drawing simple objects. This section begins our quest into visual realism: how can we simulate a 3D world in the computer and then compute images that render out like photographs?

3D Productions are built up on three basic building blocks: modeling, rendering and animation. The first two blocks—modeling and rendering—form the basis for this section. Together they determine the final displayed image.

In Chapters 5 and 6, we will study the basics of modeling and rendering, and in Chapter 7, we will introduce you to some of the more advanced concepts in the field. Chapter 8 is entirely devoted to Maya, the most popular 3D tool used in the graphics industry.

By the end of the section, you will be able to define a 3D world and render it on the computer to produce stunning imagery. So hold your breath and get ready for the 3D roller coaster.

Chapter 5

3D Modeling

The first and most critical pillar of 3D graphics is modeling. Modeling is the process of creating a 3D model in the computer.

In lay terms, a model is a representation of a concrete or abstract entity. This representation can be of various kinds. Quantitative models use equations to represent and describe system behavior; organizational models use hierarchies to represent classification schemes. A CG model refers to the geometrical representation of the entity. The purpose of the model is to allow people to visualize the structure of the entity being modeled.

When we model an object on the computer, we tell the computer about the shape, spatial layout, and connectivity of components that compose the object. In the last section, we saw examples of how to create simple 2D models such as polygons and circles. In this chapter, we expand our horizons to a 3D world. We will learn the process of viewing three-dimensional worlds and how to create, compose, and transform models in this world.

In this chapter, you will learn the following concepts

- Representation of the 3D system
- 3D math: vectors, matrices and transformations
- Creating models of
    - primitive shapes
    - generic shapes
- Viewing the 3D world
- Creating hierarchical models using transformations

We begin our discussions by first extending our knowledge of the 2D system to 3D.

5.1 The 3D System

We can relate to a three-dimensional space because we see our own world in 3D. Not only do objects in our world have length and height, as they do in the 2D space, but they also have depth associated with them. They can also be located closer or farther away from us.

In order to represent a 3D world, we need to extend our coordinate system into three dimensions. Every point in the 3D world is located using three coordinate values instead of 2. In order to define points with three coordinates, we define a third axis, typically called the *z*-axis. The *z*-axis is perpendicular to the *x*- and *y*- axis (i.e., it is at 90 degrees with respect to the two axes - also referred to as orthogonal). All three axes meet at the origin defined as (0,0,0) as shown in Fig.5.1. This is our 3D world coordinate system. In this book, we follow a right-handed coordinate system, which means that the positive *z*-axis points out of the screen and towards us.

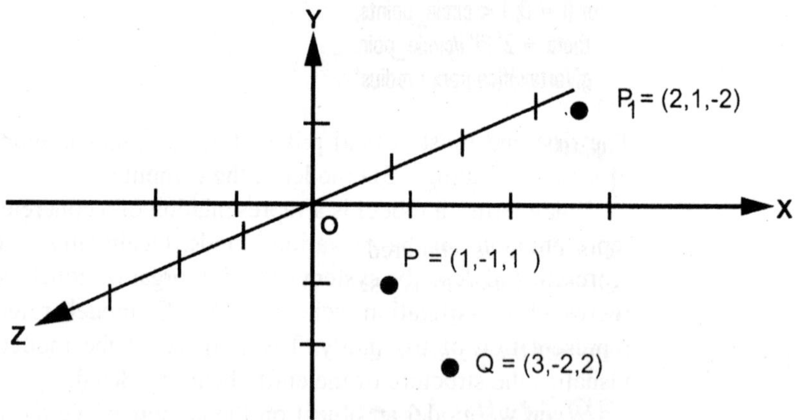

Fig.5.1: The three dimensional coordinate system

All points and vector representations that we learned about in chapter 2 can be extended to 3d by adding a third component, *z*. Points in 3D space are identified by a triplet of (*x*,*y*,*z*) values. The vector notation for this triplet is represented as: $\begin{bmatrix} x \\ y \\ z \end{bmatrix}$

In Chapter 1, we saw how objects are defined by the key (2D) vertices defining its shape. 3D vertices define the shape of on object in 3 Dimensions. The OpenGl function to define a vertex point (*x*,*y*,*z*) in 3D is

```
glVertex3f(x,y,z)
```

Using this function, we can modify our circle-drawing function from Chapter 1, to the 3D world. In the following code, we define the points of the circle in the *x*-*z* plane (not the *x*-*y* plane).

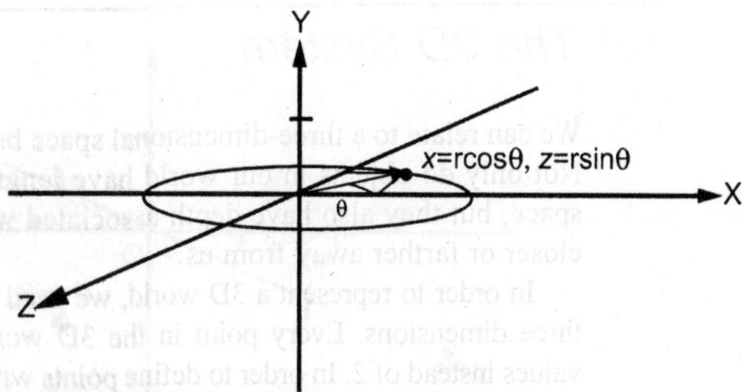

Fig. 5.2: Circle on the XZ plane

```
void MyCircle3f(GLfloat centerx, GLfloat centery, GLfloat centerz, GLfloat radius){
  GLint i;
  GLdouble theta;
  glBegin(GL_POINTS);
  for (i = 0; i < circle_points; i++) {
    theta = 2*PI*i/circle_points;
    glVertex3f(centerx+radius*cos(theta), centery, centerz+radius*sin(theta));
  }
  glEnd();
}
```

The circle is centered at the point: (centerx, centery, centerz). The function glBegin function is used to define the primitive being drawn. In this case, we merely draw points along the circumference of our circle. How would you connect the points to draw a connected circle?

3D Math: Vectors

A vector in 3D has the same semantics as in 2D – it can be thought of as a displacement from one point to another. Consider two points *P*(1,-1,1) and *Q*(3,-2,2) as shown in Fig.5.3. The displacement from *P* to *Q* is the vector

$$V = (2,-1,1) = \begin{bmatrix} 2 \\ -1 \\ 1 \end{bmatrix}$$

calculated by subtracting the coordinates of the points individually. This means that to get from point *P* to *Q*, we shift right along the *x*-axis by 2 units, down the *y*-axis by one unit and outward by one unit along the *z* axis.

Any point P_1 with coordinates (*x,y,z*) corresponds to the vector $\mathbf{V}_{p1}$, with its head at (*x,y,z*) and tail at (0,0) as shown in. That is, a 3D point is essentially a vector with its tail at the origin (0,0,0). A unit vector is a vector that has a magnitude of one unit. For example, the unit vector along the *z*-axis is: $\begin{bmatrix} 0 \\ 0 \\ 1 \end{bmatrix}$

In Example 5.1, we draw unit vectors along the positive *x-,y-,z*-axes in red

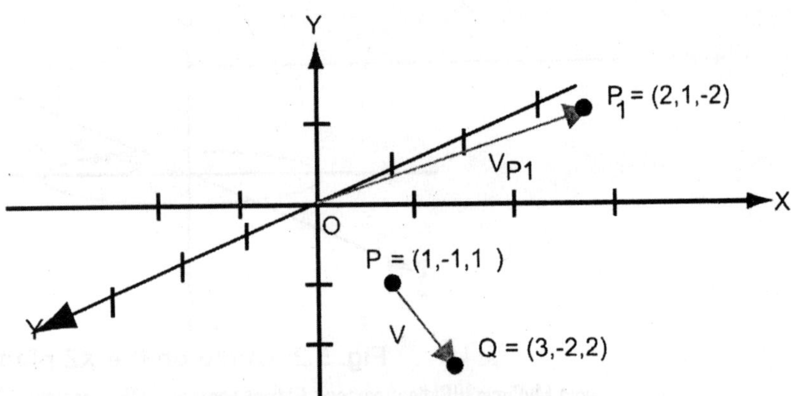

Fig. 5.3: Vectors in 3D Space

green and blue colors respectively. The code to draw the axes is as follows:

```
//X Axis
glBegin(GL_LINES);
        glColor3f (1.0, 0.0, 0.0);
        glVertex3f(0,0.,0.);
        glVertex3f(1,0,.0);
glEnd();
//Y axis
glBegin(GL_LINES);
        glColor3f (0.0, 1.0, 0.0);
        glVertex3f(0.,0.,0.);
        glVertex3f(0.,1,0);
glEnd();
//Z axis
glBegin(GL_LINES);
        glColor3f (0.0, 0.0, 1.0);
        glVertex3f(0.,0.,0.);
        glVertex3f(0.,0.,1.);
glEnd();
```

We then call the circle-drawing routine to draw two circles with unit radii, one centered (0,0.5,0) and the other at (0,0,0):

```
glPointSize(2.);
MyCircle3f(0.,0.,0.,1);
MyCircle3f(0.,0.5,0.,1);
```

The circle-drawing routine has additional code to vary the colors of the point based on its location in space. The output displays the three unit axes and the two circles. The entire code can be found in *Example5_1/Example5_1.cpp*. This example

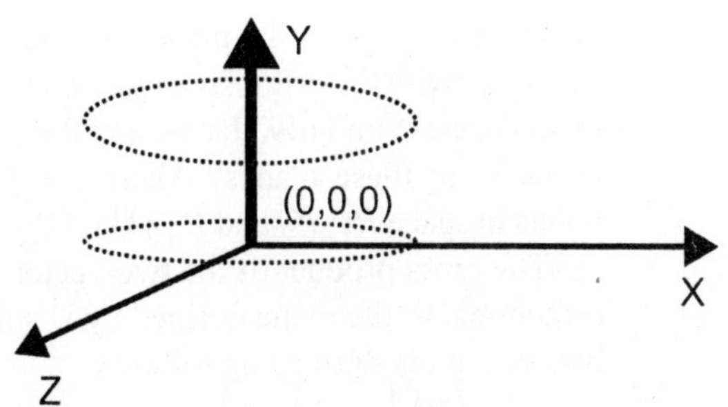

Fig. 5.4: Two circles in 3D Shape

contains lines of code that you do not understand as of yet – specifically with regards to setting up the 3D world. Do not worry; by the end of this chapter, it will all start to make sense.

Operations with Vectors

The vector math that we learned in Chapter 2 can be easily extended to 3D space. The rules for vector addition, subtraction, and multiplication rules remain the same– since all of these operations are component wise operations, there is just one more component added to the mix.

For example, if $\mathbf{V}_1$ is the vector $(x1,y1,z1)$ and $\mathbf{V}_2$ is the vector $(x2,y2,z2)$, then $\mathbf{V}_1 + \mathbf{V}_2$ is defined as:

$$\begin{bmatrix} x1 \\ y1 \\ z\,1 \end{bmatrix} + \begin{bmatrix} x2 \\ y2 \\ z\,2 \end{bmatrix} = \begin{bmatrix} x1 + x2 \\ y1 + y2 \\ z\,1 + z\,2 \end{bmatrix}$$

As shown in Fig.5.5, the semantics of the addition remains the same as in 2D. Subtraction is similarly extended.

Fig.5.5: Vector addition in 3D space

Multiplying a 3D vector **V** in space by a scalar *s* essentially scales the vector's magnitude, but does not change its direction. The process of scaling a vector such that its magnitude is of unit length is known as *normalization*. There is one vector operation not found in 2D but crucial to 3D math – that of the vector cross product. A cross product of two vectors is best understood in terms of planes and normal vectors.

A plane is a flat surface, much like a piece of cardboard or paper. Any set of points in a 2D world (by definition) lies on the same 2D plane. Points on the same plane are called *coplanar*. A plane is identified in the 3D world by a point on the plane and a vector perpendicular to the plane. This vector is referred to as the *normal vector* to the plane. For example, the *x*- and *z*-axes define a plane (the *x-z* plane) with the origin as a point on the plane and the *y*-axis as the normal vector to this plane. Similarly, the *x-z* and the *y-z* axes define planes. Can you identify the normals for these planes? Alternatively two vectors in space (defined by three points in space P_1, P_2 and P_3) **U** = (P_3 - P_1) and **V**=(P_2 - P_1) also define a plane.

The cross product of the two vectors **U** and **V** defines a third vector, which is orthogonal to the plane defined by **U** and **V**. Mathematically the cross product of two vectors is defined as follows:

If $\mathbf{U} = \begin{bmatrix} u1 \\ u2 \\ u\,3 \end{bmatrix} \mathbf{V} = \begin{bmatrix} v1 \\ v2 \\ v\,3 \end{bmatrix}$

Then the resultant cross product of the two vectors is

$$\mathbf{U} \times \mathbf{V} = \begin{bmatrix} u2v3 - u3v2 \\ u1v3 - u3v1 \\ u1v2 - u2v1 \end{bmatrix}$$

This resultant vector is the normal vector to the plane formed by **U** and **V** as shown in Fig.5.6.We shall use the concept of normal vectors heavily when we learn about lighting in Chapter 6. The code for calculating the cross product of two vectors and normalizing the result (creating a unit vector) can be found in under the installed directory for our source code in the file: *utils.h*. The code is shown below. We assume the vector values are stored in an array of 3 elements.

```
void crossproduct(GLfloat *u, GLfloat *v, GLfloat *uv) {

uv[0] - u[1]*v[2]-u[2]*v[1];
uv[1] - u[2]*v[0]-u[0]*v[2];
uv[2] - u[0]*v[1]-u[1]*v[0];
normalize(c);

}
```


Fig.5.6: Cross product of 2 vectors

In *Example5_2*, we define three points in space:

```
GLfloat P1[3] = {0.5,0.,0.5};
GLfloat P2[3] = {0.,0.,-0.5};
GLfloat P3[3] = {-0.5,0.,0.5};
```

The vectors, **U** and **V**, formed by these three points are defined as

```
vecsub(P1,P2,U);
vecsub(P1,P3,V);
```

where vecsub is a utility function defined in *utils.h* to calculate the difference between two points. The cross product of **U** and **V** is the vector **UV**.

```
crossproduct(U,V,UV);
```

which is the (unit) normal vector to the plane defined. (Remember, this vector just specifies the direction of interest – can you guess what the vector is?). We draw the three coplanar points, and place a fourth point it's normal vector:

```
glBegin(GL_POINTS);
        glVertex3fv(P1);
        glVertex3fv(P2);
        glVertex3fv(P3);
        glVertex3fv(uv);
glEnd();
```

An easy way to define vertices is by making a call to the OpenGl function

```
glVertex3fv
```

The function expects a pointer to an array of three floating point values, as shown in the sample code above. We then draw line primitives connecting all the points:

```
glBegin(GL_LINES);
        glVertex3fv(uv);
        glVertex3fv(P1);
glEnd();
```

The result displayed forms a triangular pyramid as shown in Fig.5.7. Yes – this is our very first 3D model. That was fun, wasn't it? Let us look deeper into the process of creating 3D models.

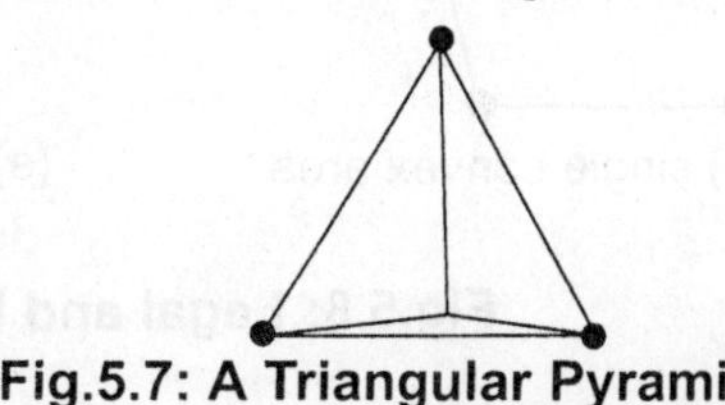

Fig.5.7: A Triangular Pyramid

5.2 3D Modeling

CG scenes depict objects of all different kinds: people, trees, flowers, fire, etc. Each one has radically different characteristics in terms of look and feel. So it should come as no surprise that there is no one standard modeling methodology—each object begs a different treatment to do full justice to the model.

The most popular and commonly used method to model objects is to use small polygons to approximate the surface. This method is popular due to ease of use, speed of display, and the abundance of algorithms to deal efficiently with polygon based models. Let us look into more detail on polygons and how they are used to model more complex shapes. In Chapter 7, we shall see some advanced techniques of representing models.

The Polygon

A polygon is much like a cutout piece of cardboard. Drawing a polygon is like playing a game of connect the dots: each dot is a vertex defining the polygon. You need at least three vertices to define the polygon. Each line of the polygon is called an *edge*.

More formally, a polygon can be defined as a set of non-crossing straight lines joining coplanar points to enclose a single convex area. Typically, most graphics packages (including openGL) support single convex area polygons. A single area means that the enclosed area should not be divided. The convex requirement means that given any two points within the polygon, you must be able to draw a straight line connecting these two points without going outside the area of the polygon, as shown in Fig.5.8.

The polygons are required to be flat (that is, to be defined in a single plane).

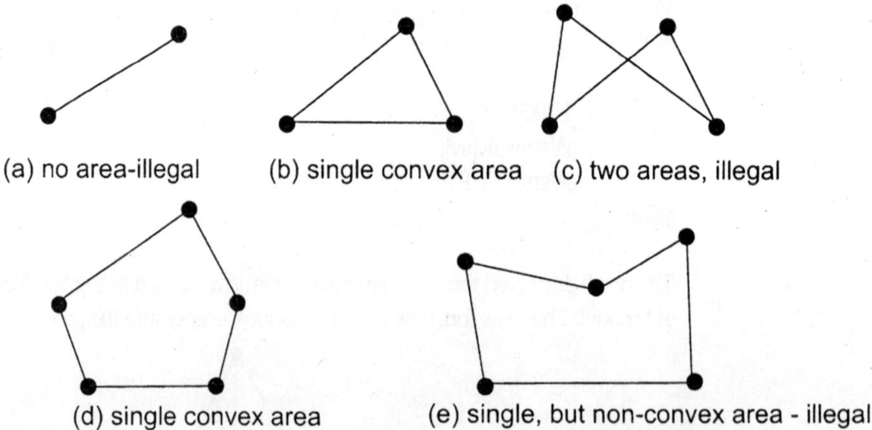

Fig.5.8: Legal and Illegal Polygons

We saw earlier that three points in space define a plane. By definition, three points also define a specific kind of polygon – the triangle. Hence, most graphics software use triangles and triangle strips to define polygonal based models. However, this approach is not essential, as we shall see when we define our primitive objects.

Front-Facing and Back-Facing Polygons

Just as a piece of cardboard has two sides to it, a polygon has two sides or faces, referred to as the front-face and the back-face. In CG, the order in which you specify the vertices of the polygon defines the orientation of the polygon. By convention, when polygon vertices are defined in counterclockwise order on the screen the polygon is said to be front facing. If you reverse the order of the vertices, the polygon is back facing: that is its back face is facing the viewer. If you turn a front facing polygon around, it will become back-facing. The face of the polygon is used extensively for defining different treatments of the two faces.

Fig. 5.9: Front- and back-facing Polygons

In *Example 5_3*, we connect the vertices of our pyramid to define triangle primitives. We use GL_POLYGON as the geometric primitive passed to glBegin, with three points defined in each block. We take care to define the polygons such that when the pyramid turns, the vertices of the face facing the camera are in counterclockwise order.

```
void myPyramid(){
        // front face, vertices defined in counterclockwise order
        glColor3f (1.0, 0.0, 0.0);
        glBegin(GL_POLYGON);
                glVertex3fv(P1);
                glVertex3fv(uv);
                glVertex3fv(P3);
        glEnd();
        // left face, vertices defined so that when this face is facing the camera, the points are in
counter-clockwise order
        glColor3f (0.0, 1.0, 0.0);
        glBegin(GL_POLYGON);
                glVertex3fv(P1);
                glVertex3fv(P2);
                glVertex3fv(uv);
```

```
        glEnd();
        //right face
        glColor3f (0.0, 0.0, 1.0);
        glBegin(GL_POLYGON);
                glVertex3fv(P3);
                glVertex3fv(uv);
                glVertex3fv(P2);
        glEnd();
        //bottom face
        glColor3f (1.0, 0.0, 1.0);
        glBegin(GL_POLYGON);
                glVertex3fv(P3);
                glVertex3fv(P2);
                glVertex3fv(P1);
        glEnd();
}
```

This function defines four polygons grouped together to form the shape of a pyramid. The polygons are colored differently for sake of identification. By default, OpenGL draws polygons as filled (with the current color).
The openGL function

```
glPolygonMode
```

can be used to define the treatment for each face of the polygons drawn.

We change the drawing mode for polygons so that we see the front faces of our pyramid as filled, and the back faces as lines.:

```
glPolygonMode(GL_FRONT, GL_FILL);
glPolygonMode(GL_BACK, GL_LINE);
```

The entire code can be found under *Example5_3/Example5_3.cpp*. When you run the program, only the front face is filled with the color red. The other faces are drawn as lines, but you can still see them.

Back-face Culling

In real life, you cannot see all sides of a solid object. You see only the sides facing you! The sides facing away from you are obstructed from view. To achieve this effect in CG, we make use of the front- and back-facing polygon concept. (And you were wondering why it was such a big deal!) The front- facing polygons are facing towards the camera and drawn into the scene. The back-facing polygons are facing away from the camera and hence are culled from the scene. This process is known as *back-face culling*.

In Fig.5.10, we show our pyramid and a sphere with all the sides visible, and

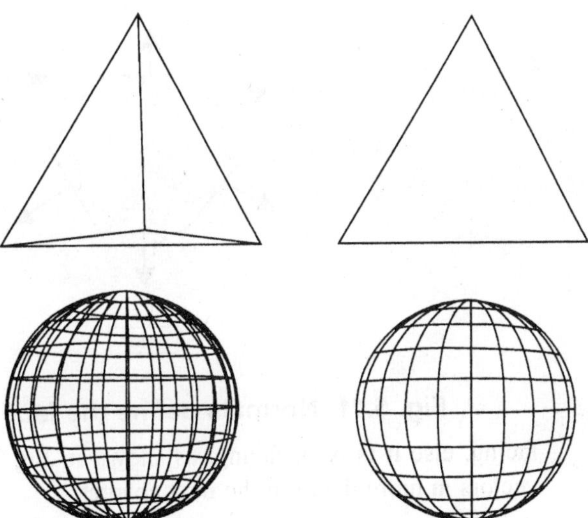

Fig.5.10: Back-face culling applied to a) pyramid, and b) sphere

then the same objects with the back faces culled from the scene, thereby generating a much more realistic image. To instruct OpenGL to discard back-facing polygons, use the command

```
glCullFace(GL_BACK)
```

You will need to enable face culling with glEnable(). If we modify the init function in *Example5_3* to be defined as

```
void init(void){
 //set the clear color to be black
 glClearColor(0.0,0.0,0.0,0.0);
 glEnable(GL_CULL_FACE);
 glCullFace(GL_BACK);
 glPolygonMode(GL_FRONT, GL_LINE);
 //glPolygonMode(GL_BACK, GL_LINE);
...
}
```

When you run the program, you will see only the pyramid's front facing polygon. The back facing polygons are culled out—exactly what we wanted. The reader is encouraged to try rotating the pyramid to see how back-face culling affects the final output. By definition, as the pyramid turns around, the faces that were back-facing become front-facing (and vice versa).

Normal vectors can also be used to determine the orientation of polygons. There is a one-to-one correspondence between the direction of the normal vector and the order in which the polygon vertices are defined. If the normal vector of the polygon points in the direction of the camera, then the polygon is front

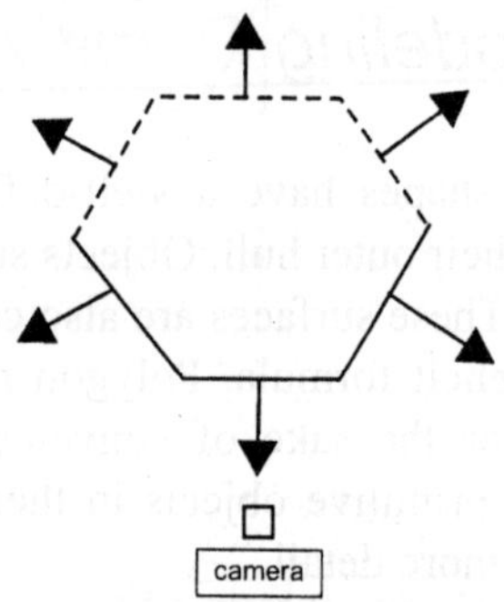

Fig. 5.11: Normals define the face of a polygon

facing, else it is back facing, as shown in Fig.5.11. We shall look into normal vectors in more detail in the next chapter.

Polygon mesh

Polygons have limited usefulness by themselves. The real power of polygons comes when they are used together in a mesh structure to make a more complex shape. A polygon mesh (referred to as *strips* in openGL) is a collection of polygons such that each edge is shared by at most two polygons, as shown in Fig.5.12.

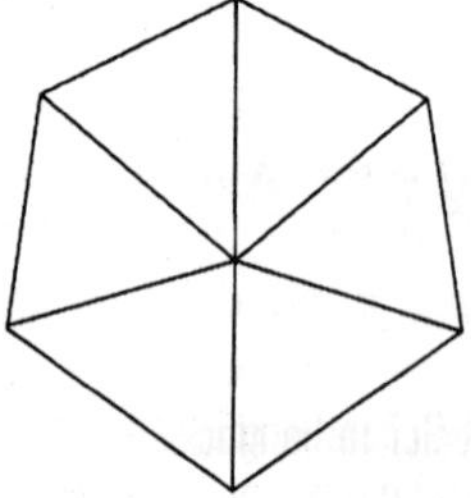

Fig.5.12: Polygon mesh - approximating a circle

In *Example 5_3*, we saw a pyramid defined by triangular polygons. Since the triangles actually share their edges, we can think of the polygons forming a solid model defined by the triangular mesh. (Ideally, triangles share edges, so the shared edges need not be redefined. For the sake of simplicity, we will let the triangles be defined individually. Since they are all connected, they effectively define a mesh.)

In fact, the outer hull of all models can be approximated by a polygonal mesh. The representation of models with just this mesh displayed is referred to as *wireframe representation*. Almost all designers start off by constructing wireframe representations of their models. Once they are satisfied with the hull of their model, they will then add lighting and material properties to render the models as a solid.

Let us look into how we can define primitive shapes using polygonal meshes.

5.3 3D Modeling Primitive Shapes

Primitive shapes have a well-defined mathematical formula that defines the shape of their outer hull. Objects such as spheres, cubes, cones, etc. fall into this category. These surfaces are also called *implicit surfaces*, since they are defined by an implicit formula. Polygon meshes can be easily used to construct these shapes. For the sake of convenience, most graphics software provide these shapes as primitive objects in their own right. Let us look into some of these shapes in more detail.

Cube

The cube model (or box) is one of the most common shapes occurring in nature. If we know the center about which the cube is based, then the length, height and depth completely define the cube as shown.

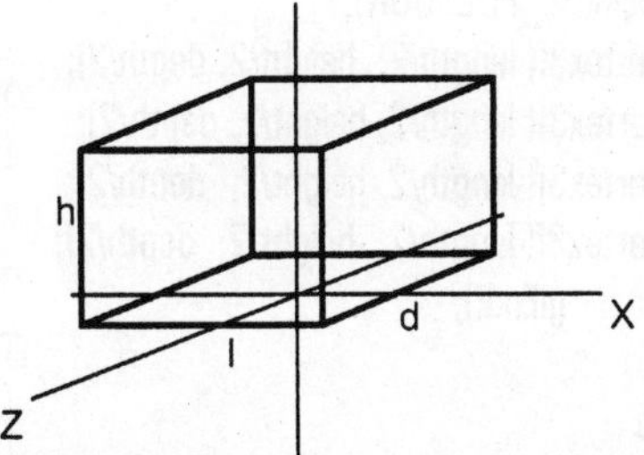

Fig.5.13: A cube defined by its three dimensions

In *Example 5_4*, we draw a cube, centered at the origin, as follows:

```
void MyCube( GLfloat length, GLfloat height, GLfloat depth){

 // base
 glBegin(GL_POLYGON);
  glVertex3f(-length/2, -height/2, depth/2);
  glVertex3f(-length/2, -height/2, -depth/2);
  glVertex3f(length/2, -height/2, -depth/2);
  glVertex3f(length/2, -height/2, depth/2);
        glEnd();

 // front face
 glBegin(GL_POLYGON);
  glVertex3f(-length/2, -height/2, depth/2);
  glVertex3f(length/2, -height/2, depth/2);
  glVertex3f(length/2, height/2, depth/2);
  glVertex3f(-length/2, height/2, depth/2);
        glEnd();
```

```
// right side face
 glBegin(GL_POLYGON);
  glVertex3f(length/2, -height/2, depth/2);
  glVertex3f(length/2, -height/2, -depth/2);
  glVertex3f(length/2, height/2, -depth/2);
  glVertex3f(length/2, height/2, depth/2);
        glEnd();
// back side face
 glBegin(GL_POLYGON);
  glVertex3f(-length/2, -height/2, -depth/2);
  glVertex3f(-length/2, height/2, -depth/2);
  glVertex3f(length/2, height/2, -depth/2);
  glVertex3f(length/2, -height/2, -depth/2);
        glEnd();
// left side face
 glBegin(GL_POLYGON);
  glVertex3f(-length/2, -height/2, depth/2);
  glVertex3f(-length/2, height/2, depth/2);
  glVertex3f(-length/2, height/2, -depth/2);
  glVertex3f(-length/2, -height/2, -depth/2);
        glEnd();


// top
 glBegin(GL_POLYGON);
  glVertex3f(-length/2, height/2, depth/2);
  glVertex3f(length/2, height/2, depth/2);
  glVertex3f(length/2, height/2, -depth/2);
  glVertex3f(-length/2, height/2, -depth/2);
        glEnd();

}
```

The built-in GLUT function

```
glutSolidCube
```

also generates a cube using polygons. It accepts the length of the cube as input and draws a cube with equal sides, centered at the origin. The Display routine in *Example5_4/Example 5_4.cpp* makes calls to both cube drawing routines, translating both of them for easier viewing:

```
void Display(void)
{
        glClear (GL_COLOR_BUFFER_BIT);
```

```
        glColor3f (1.0, 1.0, 1.0);
        glTranslatef(-1.,0.,0.);
        MyCube(1.,1.,1.);
        glTranslatef(2.,0.,0.);
        glutWireCube(1.);
        glFlush();
}
```

Sphere

Spheres are the most symmetrical shape occuring in nature. For this reason, they are also the simplest to define. To completely define a sphere, we need to define its radius, *R*, and the location of its center, *O*. If we center the sphere around our world origin, then any point *P* on the surface of the sphere can be defined by the angles θ and ϕ that it subtends at the origin as shown in Fig.5.14.

$$P=(R\cos\theta\cos\phi,\ R\sin\theta,\ R\cos\theta\sin\phi)$$

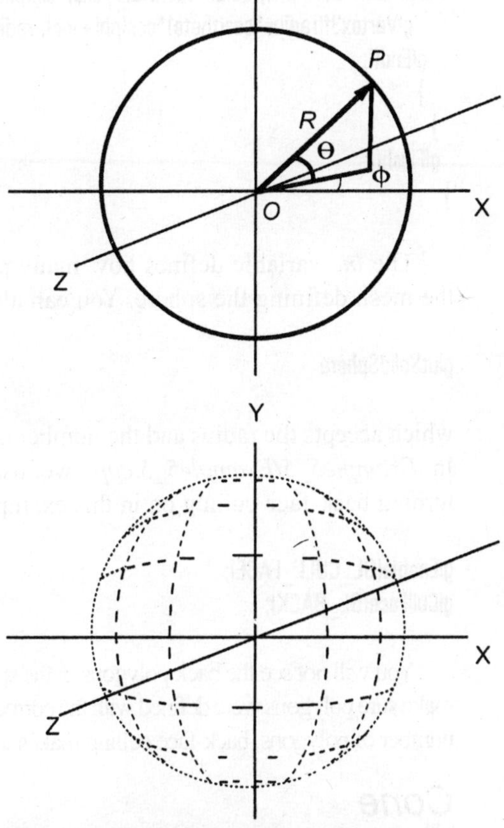

Fig.5.14: Approximating a sphere

If we incrementally advance angle θ from -90 to 90 degrees (-PI to PI radians) and ϕ from 0 to 360 degrees (0 to 2PI radians), we can define a set of vertex points approximating the surface of the sphere. These vertex points can then be connected into a polygonal mesh to define the outer hull of the sphere. The closer the intervals, the more accurate our model will be. The following shows the subroutine to calculate the polygons of the sphere:

```
void MySphere(GLfloat radius){

  GLdouble inc - PI/12;
  GLdouble theta, phi;
  bool even - true;
  for (theta--PI/2; theta<(PI/2-inc);theta+-inc){
   for (phi - 0; phi < 2*PI;phi+-inc) {
   glBegin(GL_POLYGON);
    glVertex3f(radius*cos(theta)*cos(phi), radius*sin(theta),radius*cos(theta)*sin(phi));
    glVertex3f(radius*cos(theta+inc)*cos(phi), radius*sin(theta+inc),radius*cos(theta+inc)*sin(phi));
                                                        glVertex3f(radius*cos(theta+inc)*cos(phi+inc),
radius*sin(theta+inc),radius*cos(theta+inc)*sin(phi+inc));
    glVertex3f(radius*cos(theta)*cos(phi+inc), radius*sin(theta),radius*cos(theta)*sin(phi+inc));
   glEnd();
   }
  }
  glFlush();
}
```

The *inc* variable defines how many polygon slices and stacks we define for the mesh defining the sphere. You can also use the GLUT built-in function:

```
glutSolidSphere
```

which accepts the radius and the number of slices and stacks desired as arguments. In *Example5_5/Example5_5.cpp*, we use both calls to generate spheres. Try turning back-face culling on in this example by making calls to the functions

```
glEnable(GL_CULL_FACE);
glCullFace(GL_BACK);
```

You will not see the back polygons in the sphere that we defined. Yes, we did take care to make sure polygons were defined with the correct orientation. In the case of objects with a large number of polygons, back-face culling makes a big difference in the clarity of the image.

Cone

A cone can be thought of as having a circular base and a conical surface. The

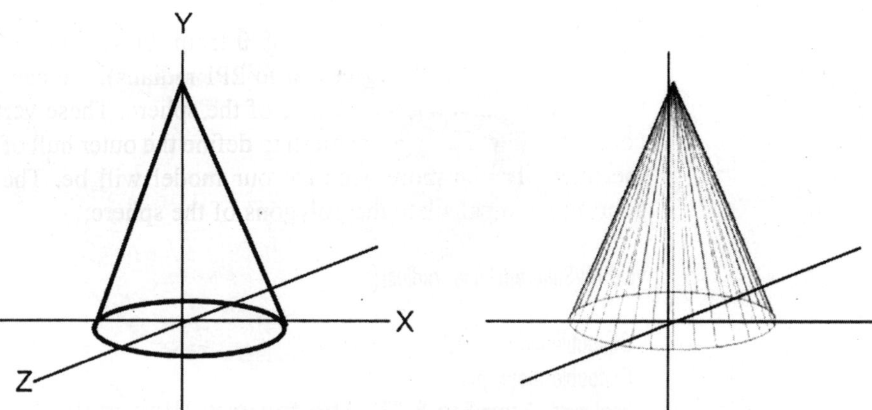

Fig.5.15: Approximating a cone

base of the cone can be defined as a circle on the *x-z* plane: the math is similar to the way we constructed a circle in Chapter 1. We approximate the conical surface of the cone by drawing a set of lines connecting the tip of the cone (a point along the *y*-axis) to the vertices along the circular base. The following code shows how to draw a cone (upper conic only) in a 3D world using openGL. The cone is assumed to be centered about the origin.

```
void MyCone(GLfloat radius, GLfloat height){
 GLint i;
 GLdouble theta, ntheta;
 for (i = 0; i < circle_points; i++) {
          glBegin(GL_POLYGON);
                    theta = (2*PI*i/circle_points);
                    ntheta = (2*PI*(i+1)/circle_points);
                    glVertex3f(radius*cos(theta), 0, radius*sin(theta));
                    glVertex3f(0, height, 0);
                    glVertex3f(radius*cos(ntheta), 0, radius*sin(ntheta));
          glEnd();

 }
```

In *Example 5_6,* draw this cone, as well as the glut primitive:

```
glutSolidCone()
```

The entire code can be found in *Example5_6/Example5_6.cpp*. Some other primitive shapes that glut provides support for are the disk, isohedron, and teapot. The teapot was first modeled in 1975 using curve and surface modeling techniques. It has now become an icon for CG, and hence is included in the primitives list. The call to generate a teapot in glut is

```
glutWireTeapot(int size)
```

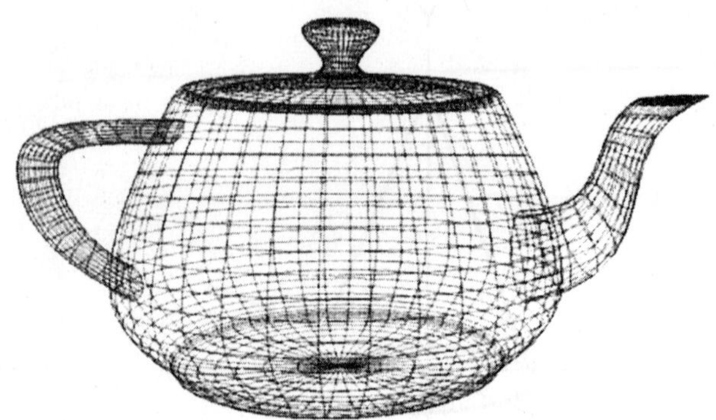

Fig.5.16: The famous Utah teapot

Fig5.16 shows the teapot image generated using this function.

5.4 3D Modeling: Generic Shapes

So far, we have been discussing primitive objects that have a well-defined mathematical formula to define their hull. What about objects that are more generic in their representation? Surely, we cannot hope to find equations to represent every possible object that occurs in nature.

For any kind of generic shape, we can always define a polygonal mesh with sufficient complexity to approximate its hull. Chapter 7 details how Nurb surfaces can be used to approximate hulls with greater accuracy and efficiency. The reader is encouraged to imagine some generic 3D models and to try modeling them using polygons, as discussed. It's not an easy task! How would you go about defining the 3D vertices to model a tree, for example? Defining the vertices of the polygons and/or curves in a 3D space can be a nerve racking experience if you don't have the necessary mathematical tools. For this reason, a plethora of modeling tools exist that will allow you to model your objects in 3D. We shall talk more about Maya, the most widely used modeling tool, in Chapter 8.

Just like image information, model information can be saved in files. For this chapter, we make use of model files to illustrate how generic objects can be modeled using polygons.

Model files

What is stored in a model file? If the model is a primitive, then the key parameters defining the primitive are saved. If it is a generic model, then information about the (polygonal or cubic) primitives used to construct it are saved.

There are many formats in which the model information can be stored. Maya stores models in its own propriety format. For this book, we shall use an open format called the *VRML* format. You can find many VRML models available for free over the Internet. Indeed, we encourage you to go ahead and download VRML model files yourself to experiment with. VRML files store polygonal information as follows:

For each model in the file:

First, all the coordinates of the vertex points are listed as *(x,y,z)* triplets. Then the index of each vertex in this list is used to specify the vertices forming the polygonal faces of the model, with a -1 as the delimiter. For example, to draw two triangles that form the faces of a unit square (Fig.5.17), the vertices would be listed as

```
0. 0. 0. // vertex 0
1. 0. 0., // vertex 1
1. 1. 0., // vertex 2
0. 1. 0. // vertex 3
```

and the indices forming the faces would be listed as

```
0,1,2,-1 // bottom triangle
2,3,0,-1 // top triangle
```

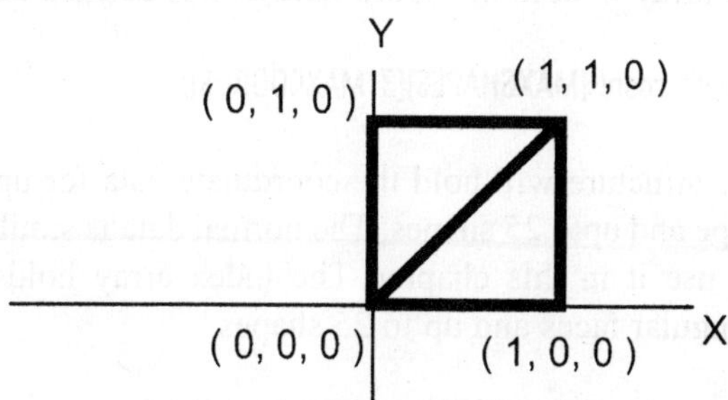

Fig.5.17: Triangular faces forming a square

Normal vectors to each polygon/vertex are defined in a similar manner. First the normal vectors are defined, and then the indices are defined determining the face to which the normal applies. Of course, we can actually calculate the normal vectors for each polygon, by computing the cross product of its defining vectors! VRML models also contain material information, which we shall not look into. A version of code to read in a very simple VRML file can be found under the directory where you installed our software, in the files: *vrml.cpp* and *vrml.h*. In these files, we define a function

```
int ReadVRML(const char *filename,
          GLfloat *coordinates, GLfloat *normals,
          GLint *indices, GLint *nindices,
          GLint *noofpoly, int MAXSHAPES, int MAXCOORDS
);
```

that will read the VRML file, and fill in the coordinate and normal vertex data. *indices* and *nindices* array hold the index values of the vertices which define the polygons and normals for this shape.

noofpoly array holds a count of the number of polygons read for each shape in the file.

MAXSHAPES determines the maximum number of shapes that will be read from the file.

MAXCOORDS indicates the maximum number of coordinates that each shape may have.

In *Example 5_7*, we read in the VRML file *robot.wrl*: a model of an Android model. The model itself consists of different shapes forming the head, legs, hands, etc of the Android. The code to read in the model is simple. We define the MAXSHAPES that we can read in to be 25 and the maximum coordinate data per shape as 3000:

```
#define MAXSHAPES 25
#define MAXCOORDS  3000;
```

The array to hold the coordinate data is defined as

```
GLfloat   coords[MAXSHAPES][3*MAXCOORDS];
```

This structure will hold the coordinate data for up to 3000 vertices (9000/3)per shape and upto 25 shapes. The normal data is similarly defined, although we will not use it in this chapter. The index array holds information for up to 3000 triangular faces and up to 25 shapes.

```
GLint   indices[MAXSHAPES][3*MAXCOORDS];
```

For each shape, both arrays hold the three coordinates/indices of each vertex/polygon in a linear one-dimensional manner. We maintain a count of the number of polygons defined by each shape

```
GLint noofpoly[25];
```

In the main code, we read in an android model as follows:

```
noofshapes = ReadVRML("..\\Models\\robot.wrl", &coords[0][0], &normals[0][0],
&indices[0][0],&nindices[0][0],
&(noofpoly[0]), 25, 3000);
```

We can draw the models in many different ways. The simplest way is to draw the triangular faces one by one using the glVertex command.

```
for (j=0;j<noofshapes;j++) {
        for (i=0;i<noofpoly[j]*3;i=i+3) {
                glBegin(GL_TRIANGLES);
        glVertex3f(coords[j][3*indices[j][i]],
                        coords[j][3*indices[j][i]+1],
                        coords[j][3*indices[j][i]+2]);
                glVertex3f(coords[j][3*indices[j][i+1]],
                        coords[j][3*indices[j][i+1]+1],
                        coords[j][3*indices[j][i+1]+2]);
                glVertex3f(coords[j][3*indices[j][i+2]],
                        coords[j][3*indices[j][i+2]+1],
                        coords[j][3*indices[j][i+2]+2]);
        glEnd();
        }}
```

An easier way to define the vertices is by making a call to the OpenGl function

```
glVertex3fv
```

Using this function, the display code can be modified to be

```
for (j=0;j<noofshapes;j++) {
```

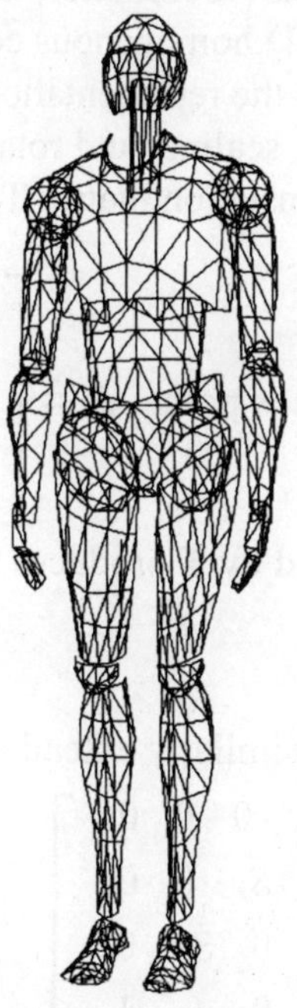

Fig.5.18: An Android, read from a VRML model

```
        for (i=0;i<noofpoly[j]*3;i=i+3) {
        glBegin(GL_TRIANGLES);
                glVertex3fv(&(coords[j][3*indices[j][i]]));
                glVertex3fv(&(coords[j][3*indices[j][i+1]]));
                glVertex3fv(&(coords[j][3*indices[j][i+2]]));
        glEnd();
}}
```

Fig.5.18 shows the output of the code. The entire code can be found in *Example5_7/Example5_7.cpp*, and the VRML file, *robot.wrl,* can be found under the *Models* directory under the installed folder for our sample code. You will need to compile and link the program with the files: *vrml.h* and *vrml.cpp*. Readers are encouraged to experiment with their own VRML models. You can download these models for free from the Internet. You may have some trouble with world coordinates of differing objects. We discuss how to modify the world coordinate space in 3D in the next section.

5.5 3D Transformations

We have seen how to create models in the 3D world. We discussed transforms in Chapter 2, and we have used some transformations to view our 3D models effectively. Let us look into 3D transformations in more detail.

Just as 2D transformations can be represented by 3 x 3 matrices using homogenous coordinates, 3D transformations can be represented by 4 x 4 matrices. 3D homogenous coordinates are represented as a quadruplet (x,y,z,W). We discuss the representation of the three most commonly used transformations: translation, scaling, and rotation.

The translation matrix $\mathbf{T}(Tx, Ty, Tz)$ is defined as

$$\mathrm{T} = \begin{bmatrix} 1 & 0 & 0 & Tx \\ 0 & 1 & 0 & Ty \\ 0 & 0 & 1 & Tz \\ 0 & 0 & 0 & 1 \end{bmatrix}$$

so any vertex point $P = \begin{bmatrix} x \\ y \\ z \\ 1 \end{bmatrix}$

transformed by $\mathbf{T}$ produces the transformed point $P' = \mathbf{T}.P = \begin{bmatrix} x + Tx \\ y + Ty \\ z + Tz \\ 1 \end{bmatrix}$

Scaling is similarly extended to be defined as

$$\mathrm{S} = \begin{bmatrix} Sx & 0 & 0 & 0 \\ 0 & Sy & 0 & 0 \\ 0 & 0 & \mathrm{Sz} & 0 \\ 0 & 0 & 0 & 1 \end{bmatrix}$$

Recall from Chapter 2 that scaling occurs about a point. This point is called the *pivot point*. The scaling matrix defined above scales about the origin. You can define models to have different pivot points by setting the transformation stack appropriately. We shall see in greater detail, on how models can be defined to have a predefined pivot point.

Rotation matrices are defined uniquely based on the axis of rotation and the pivot point. The axis of rotation is a normalized vector defining the axis in 3D space along which the rotation will occur. (A 3D vector will simply be the triplet (x,y,z) defining the direction of the axis). The 2D rotation we saw earlier is a 3D rotation about the z-axis (the vector (0,0,1)), and the world origin (0,0,0) and is defined as:

$$Rz= \begin{bmatrix} \cos(\theta) & -\sin(\theta) & 0 & 0 \\ \sin(\theta) & -\cos(\theta) & 0 & 0 \\ 0 & 0 & 1 & 0 \\ 0 & 0 & 0 & 1 \end{bmatrix}$$

Similarly the rotation matrix about the x-axis is defined to be

$$Rx= \begin{bmatrix} 1 & 0 & 0 & 0 \\ 0 & \cos(\theta) & -\sin(\theta) & 0 \\ 0 & \sin(\theta) & \cos(\theta) & 0 \\ 0 & 0 & 0 & 1 \end{bmatrix}$$

and about the y-axis as:

$$Ry= \begin{bmatrix} \cos(\theta) & 0 & \sin(\theta) & 0 \\ 0 & 0 & 0 & 1 \\ \sin(\theta) & 0 & \cos(\theta) & 0 \\ 0 & 0 & 0 & 1 \end{bmatrix}$$

Just as with 2D transforms, 3D transforms can be composed together to provide desired effects. The order in which the transforms are applied is important. Recall from our previous section that the OpenGL commands for matrix operations

void glLoadIdentity(void)

sets the current matrix to be the 4×4 identity matrix.

glLoadMatrix4f(GLfloat *m)

sets the matrix passed in to be the current transformation matrix.

glMultMatrix4f(Glfloat *m)

multiplies the current matrix by the matrix passed in as the argument.

The argument for all these commands is a vector of 16 values (m_1, m_2,..,m_{16}) that specifies a matrix **M** as follows:

$$M = \begin{bmatrix} m1 & m2 & m3 & m4 \\ m5 & m6 & m7 & m8 \\ m9 & m10 & m11 & m12 \\ m13 & m14 & m15 & m16 \end{bmatrix}$$

Additionally, OpenGL provides three utility functions for the most commonly used transformations.

```
glTranslatef(Tx,Ty,Tz)
```

multiplies the current transformation matrix by the matrix to move an object by (Tx,Ty,Tz) units

```
glRotatef(theta, x,y,z)
```

multiplies the current transformation matrix by a matrix that rotates an object in a counterclockwise direction by theta degrees about the axis defined by the vector (x,y,z)

```
glScalef(Sx,Sy,Sz)
```

multiplies the current transformation matrix by a matrix that scales an object by the factors, (Sx,Sy,Sz).

All three commands are equivalent to producing an appropriate translation, rotation, or scaling matrix and then calling glMultMatrixf() with that matrix as the argument. Let us apply some transformations to the famous pyramid that we created earlier. In *Example5_8*, we rotate the pyramid about the *y*-axis and use double buffering to display the rotation as an animation.

```
glRotatef(0.1,0.,1.,0.);
//glRotatef(0.1,0.,0.,1.);
glutPostRedisplay();
```

The entire code can be found in *Example5_8/Example5_8.cpp*. Readers should experiment rotating about both the *x*- and *z*-axes to see what results they get So now that we have covered some of the basics of 3D modeling, you are probably asking: how are we defining the 3D world? What are its extents? How is this world being mapped onto the 2D screen? Let us take a closer look at how this process is accomplished.

In Chapter 2, we saw how to map a two-dimensional world onto the two dimensional screen. The process was a simple mapping of 2D world coordinates onto the 2D viewport of the window. Model positions are first clipped against the 2D world coordinates and then mapped into the viewport of the window for display.

5.6 Viewing in 3D

The solution to transforming a 3D world into a 2D plane is a little more complex and is accomplished by the use of projections. Projections provide the transforms needed to map 3D points onto a 2D plane, as shown in Fig.5.19. To determine the projection, we need to define a view volume of the world (the world coordinates), a projection that defines the mapping of the view volume onto a projection plane, and the viewport for final display. Conceptually, objects are first clipped against the 3d view volume, projected, and then finally mapped into the specified viewport.

Whew! That was a mouthful. Let us see if we can understand this process by way of a real-world analogy. The essential goal in 3D CG is to give the viewer the impression that he is looking at a photograph of a three-dimensional scene, much the same way we photograph our 3D world onto a 2D film. 3DCG simulates the process of the real-world camera to create 3d images. Let us first understand how a real camera records an image of our real live 3D world by using the example of a simple pinhole camera. We will then extend this discussion to see how CG simulates this process.

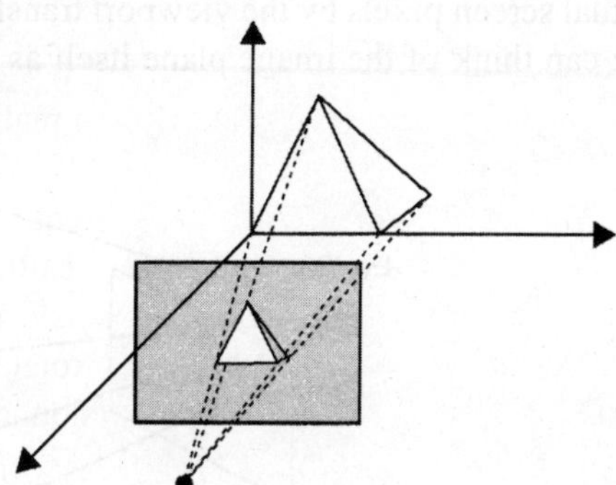

Fig.5.19: Projecting a 3D object onto a plane

Pinhole Camera

Fig.5.20 shows a pinhole camera. The camera is a simple box with photographic film F at the back and a small hole H that allows light in when opened.

Consider what happens when we take a photograph of object X. We open hole H for a fraction of a second. A ray of light hits X at point P, passes through H and hits the film at P', causing the film to be exposed at this point. The point P' forms the image of point P on the negative. All points on the object are mapped onto the film in this fashion. Another object Y, of the same size but

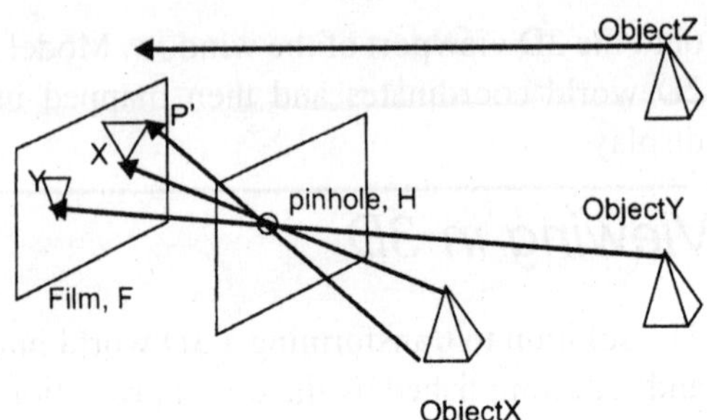

Fig.5.20: Pinhole Camera

further away, gets mapped on as Y' and is actually smaller in size. Objects such as Z, whose intersecting rays of light do not pass through H, are not photographed. Notice that the image formed is inverted! The classic computer graphics version of the camera places the film in front of the pinhole. This placement ensures among other things, that the image is right side up. We rename the pinhole as the *eye*, *viewpoint*, or *camera position* and the film as the *image plane*, as shown in Fig.5.21.

The process of filming the scene remains very much the same. Rays (called *projectors*) are generated from the viewpoint to the objects. Objects for which these rays intersect the image plane will be processed and displayed on screen. Others such as Object Z, will be clipped from the image.

The image plane itself can be thought of as being composed of a two-dimensional grid containing color information that is then mapped onto the actual screen pixels by the viewport transformation. For the sake of convenience, we can think of the image plane itself as being composed of pixels.

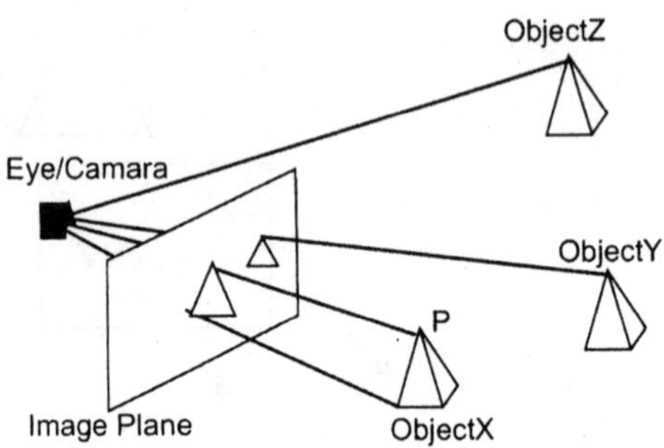

Fig.5.21: Computer Camera

The CG Camera

The steps needed to generate the CG image can be defined analogously to the steps we would take a photograph in the real world. We need to define the following:

1. **Viewing Transformation**

   The viewing transformation sets the viewing position and is analogous to positioning and aiming a camera.

2. **Object/Modeling Transformation**
The modeling transformations are used to position and orient the model within the world. This process is akin to arranging the scene to be photographed into the desired composition. For example, you can rotate, translate, or scale the model-or perform some combination of these operations. We saw how to do this in a 2D world. In 3D, the concept is enhanced to include a third dimension defined by the z-axis.

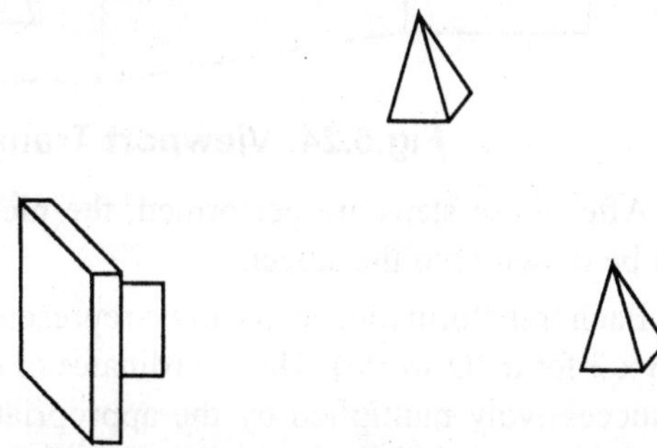

Fig.5.22: Modeling and Viewing Transformation

3. **Projection Transformation**
Specifying a projection transformation is akin to choosing a camera lens or adjusting the zoom. This transformation determines the extents of the world in view, also called the viewing volume. In addition, the projection transformation also determines the location and orientation of the image plane and how objects in the scene are clipped and projected onto this plane. This process determines the world coordinates of the scene.

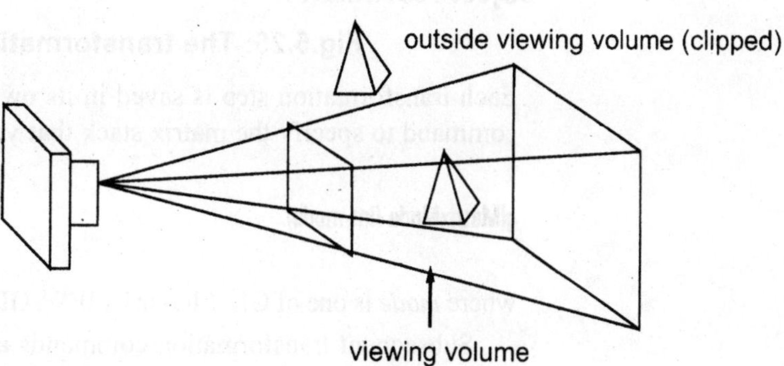

Fig.5.23: Projection Transformation

4. **Viewport transformation**
The viewport transformation determines the shape of the available screen area into which the scene is mapped and finally displayed. This process is analogous to determining how large the final

photograph will be. This step is same as in the 2D world - when we finally map the world coordinates onto the display.

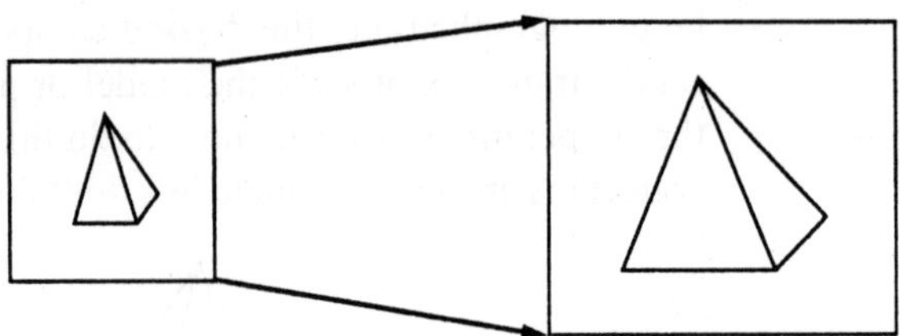

Fig.5.24: Viewport Transformation

After these steps are performed, the picture can be snapped, i.e., the scene can be drawn onto the screen.

Each transformation step can be represented by a 4 × 4 matrix M (as opposed to 3 x 3 for a 2D world). The coordinates of each vertex point P (in object space) is successively multiplied by the appropriate transformation matrix to achieve the final position P' (in world space) as shown in Fig.5.25.

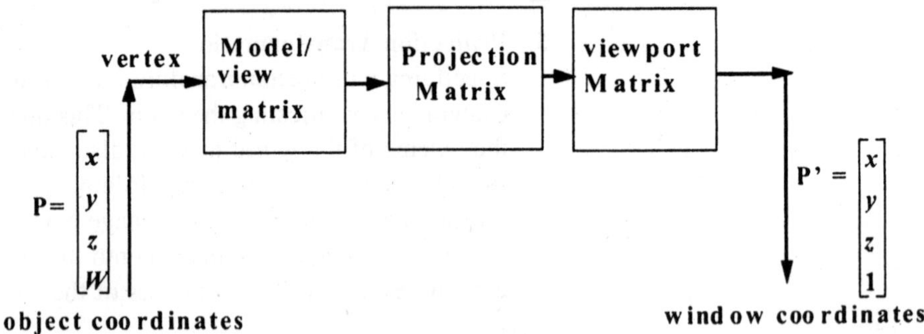

Fig.5.25: The transformation sequence

Each transformation step is saved in its own matrix stack. We saw the OpenGl command to specify the matrix stack that you wish to operate on is

```
glMatrixMode (int mode)
```

where *mode* is one of GL_MODELVIEW, GL_PROJECTION, or GL_TEXTURE.

Subsequent transformation commands affect the specified matrix stack. We saw in earlier examples how to set the ModelView matrix stack and modify the transforms applied to the models in the scene. We also saw how viewport transformations are accomplished by the OpenGL command, glViewport.

Note that only one stack can be modified at a time. By default, the ModelView stack is the one that's modifiable.

Let us look at each transformation stack one at a time. We begin with model transformations since we have already seen it before, and it's an intuitive place to begin our discussions.

Object/Model Transformations

We can move, resize, or orient the models in our 3D world differently to create a composition or scene. We saw in Chapter 2, that these transformations are called object transformations or model transformations. There is an alternative way to think about model transformations: moving the camera in one direction will have the same effect as moving all the models in the scene in the opposite direction! For this reason, viewing and modeling transformations are inextricably related in OpenGL and are in fact combined into a single modelview matrix. You should find a natural approach for your particular application that makes it easier to visualize the necessary transformations, and then write the corresponding code to specify the matrix manipulations.

You must call glMatrixMode() with GL_MODELVIEW as its argument prior to performing modeling or viewing transformations. In *Example 5_9*, we modify the ModelView matrix to move our pyramid model back in z- by 4 units, scale it up and constantly keep rotating it around the y-axis.

```
glLoadIdentity ();
// translate pyramid back by 4 units, and down by 1 unit
glTranslatef(0.,-1.,-4);
// scale pyramid
glScalef(2.,2.,1.);

// rotate the pyramid
glRotatef(ang,0.,1.,0.);

if (ang >=360) ang = 0; else ang++;
// draw the pyramid
myPyramid();
glFlush();
glutPostRedisplay();
```

Which transform is being applied first? What will happen if we change the order of the transforms? The entire code can be found in *Example5_9/Example5_9.cpp*.

Viewing Transformation

A viewing transformation changes the position and orientation of the viewpoint itself. Essentially, the viewing transformation repositions the camera and the direction it is looking. As discussed earlier, to achieve a certain scene composition in the final image or photograph, you can either move the camera or move all the objects in the opposite direction. In *Example5_9*, the translation routine can be thought of as actually moving the camera position by 4 units towards the object, instead of the object being moved back. Readers are encouraged to try experimenting with this approach of transforming the scene.

Remember that transformations are applied in the reverse order of the commands specified. Viewing transformation commands must be called before any modeling transformations are performed. This will ensure that the viewing transformations are applied to all the models after the model transforms.

You can manufacture a viewing transformation in OpenGL using different techniques. One is using the standard glTranslate and glRotate functions. A more popular method is to use the Utility Library routine gluLookAt() to define a line of sight.

```
void gluLookAt(GLdouble eyex, GLdouble eyey, GLdouble eyez, GLdouble centerx, GLdouble centery,
GLdouble centerz, GLdouble upx, GLdouble upy, GLdouble upz);
```

This call defines a viewing matrix such that the desired viewpoint is located at (eyex, eyey, eyez). The centerx, centery, and centerz arguments specifies a point in the scene that is being looked at. The upx, upy, and upz arguments indicate which direction is up (that is, the direction from the bottom to the top of the viewing volume). This matrix is then multiplied to the current modelview matrix. In the default position, the camera is at the origin, is looking down the negative z-axis, and has the positive y-axis as straight up. This is the same as calling

```
gluLookat (0.0, 0.0, 0.0, 0.0, 0.0, -100.0, 0.0, 1.0, 0.0);
```

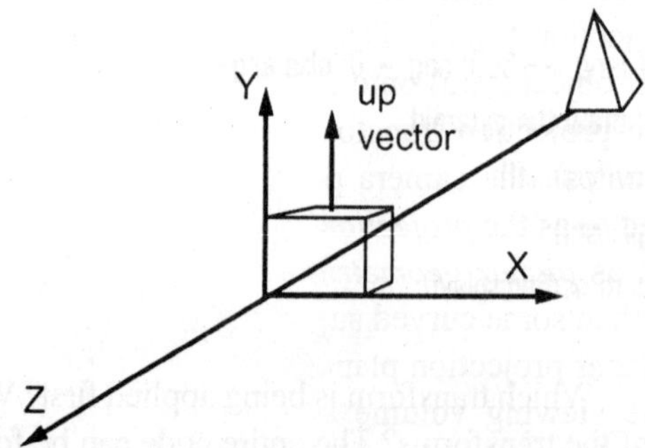

Fig.5.26: Camera position using gluLookAt

The following code from *Example5_10/Example5_10.cpp* moves the viewpoint closer to and farther from the pyramid, as the pyramid continues its rotation about the *y*-axis. Notice that we call the viewing transform before calling the object transforms. The matrix stack is specified in the init() function.

```
GLfloat zdist = 3., zmove=1;
void Display(void)
{
        glutSwapBuffers();
        glClear (GL_COLOR_BUFFER_BIT);
                glLoadIdentity ();

        gluLookAt(0.,0.,eyez, 0,0,-100,0.,1.,0.);
        if (eyez >20) zmove = -1;
        if (eyez <3) zmove = 1;
        eyez+ =zmove*0.2;

        // rotate the pyramid
        glRotatef(ang,0.,1.,0.);
        if (ang >=360) ang = 0; else ang++;
        // draw the pyramid
        myPyramid();
        glFlush();
        glutPostRedisplay();
}
```

How would you define the transforms such that the camera rotates around the pyramid? Hint: You can use the equations we derived for points along a circle.

Projection Transformations

In general, projections transform points in an *n*-dimension coordinate system to a coordinate system with fewer than *n*-dimensions. In 3DCG we are interested in the projection from a 3D space to a 2D space—just like a shadow is a projection of a 3D object (you) onto a 2D plane (the ground).

The projection transformation is defined by a viewing volume (the world coordinates), the camera position or the viewpoint, and the image plane, also referred to as the *projection plane*. The class of projections we deal with here is known as *planar geometric projections* because the projection is onto a plane rather than some curved surface. The Omnimax film format, for example, uses a non-planar projection plane.

The viewing volume and the viewpoint determine how the 3D object is projected onto the plane. We use straight lines, called *projectors*. Projectors are drawn from the center of projection (the viewpoint) to each vertex point on the object. The points of intersection between the projectors and the image plane forms the projected image. The two basic classes of planar projections are perspective and parallel, as shown in Fig.5.27.

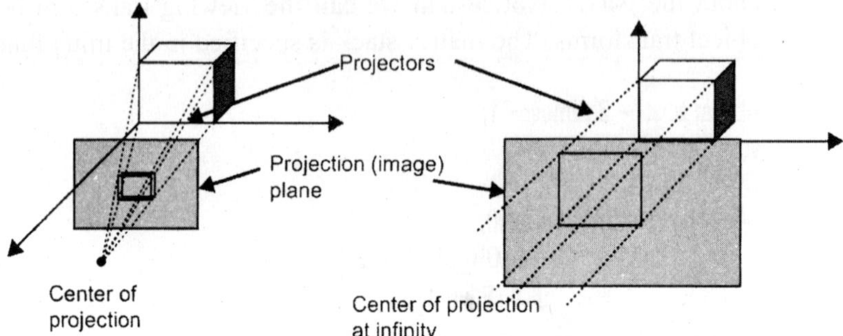

Fig.5.27: Perspective and Parallel Projection

Perspective Projection

The most unmistakable characteristic of perspective projection is perspective foreshortening: the farther an object is from the camera, the smaller it appears in the final image. This effect occurs because the viewing volume for a perspective projection is a frustum of a pyramid (a truncated pyramid whose top has been cut off by a plane parallel to its base). Objects that are closer to the viewpoint appear larger because they occupy a proportionally larger amount of the viewing volume than those that are farther away. This method of projection is commonly used for animation, visual simulation, and any other applications that strive for some degree of realism because the process is similar to how our eye (or a camera) works. The following command defines the frustrum of the viewing volume:

void glFrustum(GLdouble left, GLdouble right, GLdouble bottom, GLdouble top, GLdouble near, GLdouble far);

This function calculates a matrix that accomplishes perspective projection and multiplies the current projection matrix by it. The viewing volume is defined by the four sides of the frustum, its top, and its bottom, which correspond to the six clipping planes of the viewing volume, as shown in Fig.5.28. Essentially, the viewing volume defines the world coordinates within which our 3D objects reside. Objects or parts of objects outside this volume are clipped from the final image.

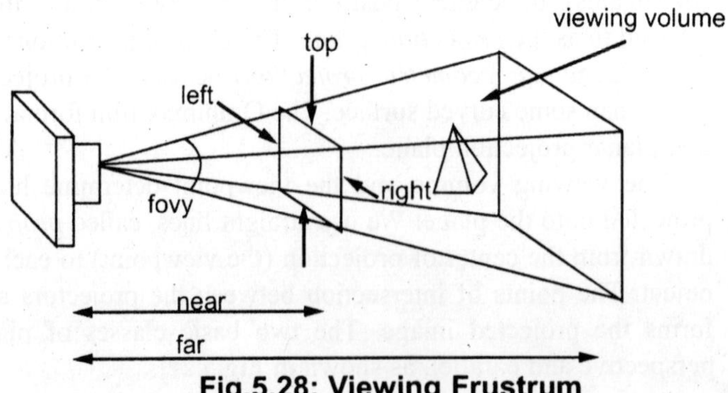

Fig.5.28: Viewing Frustrum

The projection plane is defined as a plane parallel to the base of the pyramid and centered at the viewpoint position. The following code from *Example5_11*, shows the effect of changing the frustrum parameters. Changing the left, right, bottom and top planes is like changing the viewing lens on a camera – you can view your scene with a wide angle or a telephoto lens!

```
glMatrixMode (GL_PROJECTION);
        glLoadIdentity();
        glFrustum (-1.0*lens, 1.0*lens, -1.0*lens, 1.0*lens, 1., 20);
//      gluPerspective(fang, aspect, 1., 20.0);
        glMatrixMode (GL_MODELVIEW);
        glLoadIdentity();
        gluLookAt(0.,0.,3, 0,0,-100,0.,1.,0.);
        // rotate the pyramid
        glRotatef(ang,0.,1.,0.);
        if (ang > -360) ang - 0; else ang + +;
        // draw the pyramid
        myPyramid();
        glFlush();
        glutPostRedisplay();
```

If you change the near and far clipping planes instead, you will see the pyramid gradually being clipped out of the scene. The glFrustum command is not always intuitive. An easier command to define a perspective projection is

```
void gluPerspective(GLdouble fovy, GLdouble aspect,
GLdouble near, GLdouble far);
```

fovy is the angle of the field of view in the *x-z* plane (shown in Fig.5.28) This value must be in the range [0.0,180.0].

aspect is the aspect ratio of the frustum, that is its width divided by its height.

Example5_11 can be modified to replace the command glFrustrum with the command:

```
gluPerspective(fang, aspect, 1., 20.0);
```

where the field of view (fang) of the camera changes from 30 degrees to 160 degrees to achieve the same effect as seen by glFrustrum. In general, we always use the gluPerspective function to define the perspective projection in our programs.

Parallel Projection (Orthographic)

Parallel projections are projections whose center of projection is at infinity. These projections ignore the location of the current viewpoint in their calculations. The projectors used are parallel lines.

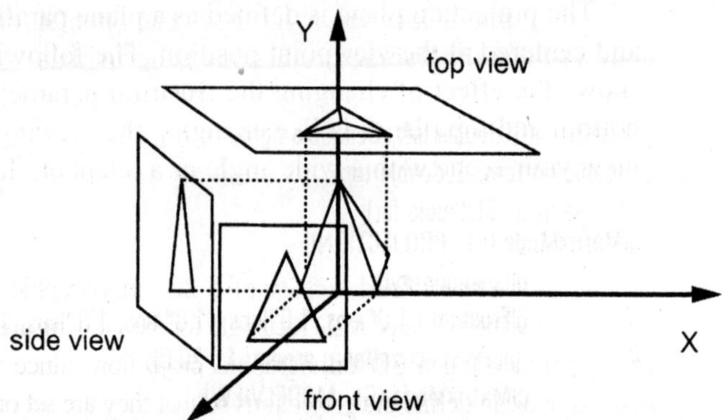

Fig.5.29: Orthographic projection: viewing volume

The output of a parallel projection is not realistic because objects retain their size no matter where they are located. It is very useful for calculating measurements and sizes of objects and is often used when modeling objects, in engineering drawing, and in blue prints of architectural plans.

Parallel projections are called *orthographic* if the direction of projection is orthogonal (at 90 degrees) to the projection plane. The most common type of ortho projections are along the *x*-, *y*- and *z*- axis (called the front, top and side views) where the projection planes are on the *y-z*, *x-z* and *x-y* planes respectively as shown in Fig.5.29. These projections are used extensively in modeling tools to enable the process of 3d modeling.

With an orthographic projection, the viewing volume (the world coordinates) is a box, as shown in Fig.5.30

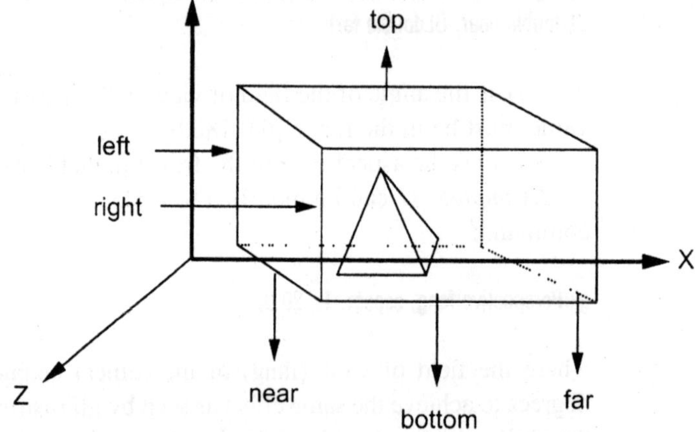

Fig.5.30: Orthographic viewing volume

Unlike perspective projection, the size of the viewing volume doesn't change from one end to the other, so distance from the camera doesn't affect how large an object appears to be. The direction of the viewpoint defines the direction of projection, but its location is ignored. The openGL command to specify the view volume for orthographic projections is defined as

```
void glOrtho(GLdouble left, GLdouble right,
GLdouble bottom, GLdouble top,
GLdouble near, GLdouble far);
```

We saw a specialized version of a 2D orthographic projection in the Chapter 2—glOrtho2d, which defined a 2D rectangle for the viewing area. In *Example 5_12*, we display the pyramid using a 3D orthographic projection. Since we don't have any animation defined, we can define the projections so that they are set only in the reshape function:

```
void reshape (int w, int h)
{
 // on reshape and on startup, keep the viewport to be the entire size of the window
        glViewport (0, 0, (GLsizei) w, (GLsizei) h);
        glMatrixMode (GL_PROJECTION);
        glLoadIdentity();
        // Set up the ortho view volume
        glOrtho(-1,1,-1,1,1.,20);
        glMatrixMode (GL_MODELVIEW);
        //Define viewing position as 5 units back in Z, and
        gluLookAt(0.,0.,5, 0,0,-100,0.,1.,0.);
        // rotate for top view
        //glRotatef(90,1.,0.,0.);
        // rotate for side view
        //glRotatef(-90,0.,1.,0.);

}
```

This code will display the front view of the pyramid. To see the top view, we need to define the direction of projection to be along the *y*-axis. This is achieved by rotating the viewing position 90 degrees about the *x*-axis by inserting the following lines of code after the gluLookat command:

```
// rotate for top view
glRotatef(90,1.,0.,0.);
```

Similarly, rotate the viewing position by 90 degrees along the *y*-axis to get a side view. The different-colored faced of the pyramid will provide you with a good indication as to where you are looking.

Viewport Transformations

We looked into details of the viewport transformation in Section 1. The viewport transformation transforms the image plane into the specified region of the window.

By default the viewport is set to the entire pixel rectangle of the window that is open. You can also use the glViewport() command to also choose a smaller drawing region; for example, you can subdivide the window to create a split-screen effect for multiple views in the same window. Try experimenting with different viewport settings within your own programs.

Transformations and matrix stacks are not only useful for defining the 3D-to-2D mapping. They prove to be an invaluable tool when constructing models as well. Let us look into this aspect of modeling before we end the chapter.

5.7 Hierarchical Modeling Using Transformations

Models are usually not constructed as monolithic blobs. They usually have a hierarchical structure: smaller models called shapes/components are used as building blocks to create higher-level entities, which in turn serve as building blocks for higher-level entities, and so on. A hierarchy enables the creation of complex objects in a modular fashion, typically by repetitive invocation of building blocks. The model hierarchy can be symbolized as a tree, with the components defined as nodes of the tree. The nodes at the top of the hierarchy are the parent objects. Parents have links to their child objects, who have further links to their children etc.

The child objects are defined in a default orientation and must be transformed in order to be correctly positioned with respect to their parent object. To accomplish this, parent objects call each child objects in the hierarchy, passing to it the appropriate geometric transforms corresponding to its scale, orientation and position. The parameters are defined within the parent's transformation space. This means that the children inherit their parent's transforms, so a link ensures that the spatial position of the child gets updated when the parent is transformed. Transforming the children has no affect on the parent.

The hierarchy of the model determines the motion definition for the model and its components. Careful thought needs to be put into constructing a hierarchy for a model. The best choice of the hierarchy is one that takes into account the movement/animation of the model within scene.

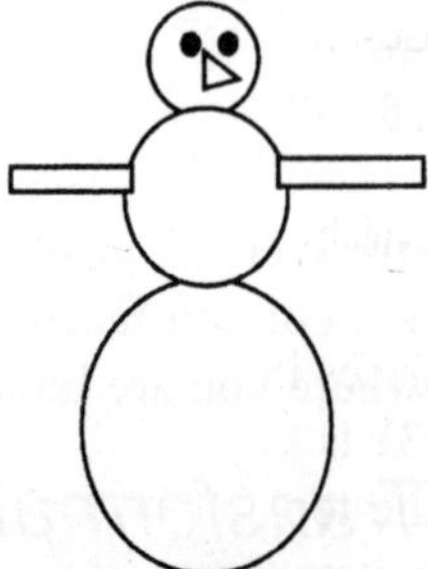

Fig.5.31: A Snowman

Consider building the model of a snowman. We shall refer to him as Snowy, as shown in Fig.5.31

Let us first consider the component objects that build up the snowman. The base, tummy, and head are of course all spheres. The hands could be modeled using cones or cylinders and the eyes using disks. The carrot shaped nose is just a simple cone.

How would we define the hierarchy of such a model? We definitely want the eyes and the nose to ride along with the face. Therefore we would define them to be children of the head object. Similarly, the hands can be grouped as children of the tummy. Now the base, tummy and head need to be parented in a way that would enable us to move them as a single unit. In order to do this, we define what we call a *Null Node*. A null node is used solely for the purpose of grouping together components in a hierarchy. It has no graphical display in our 3D world. It does have a location, which defines the pivot for any rotation and scale transforms applied to it (and hence to its children). The introduction of the null node would lead us to a hierarchical structure of the snowman as shown in Fig.5.32. Let us see how we can use matrix stacks to define the hierarchical model of the snowman using OpenGL.

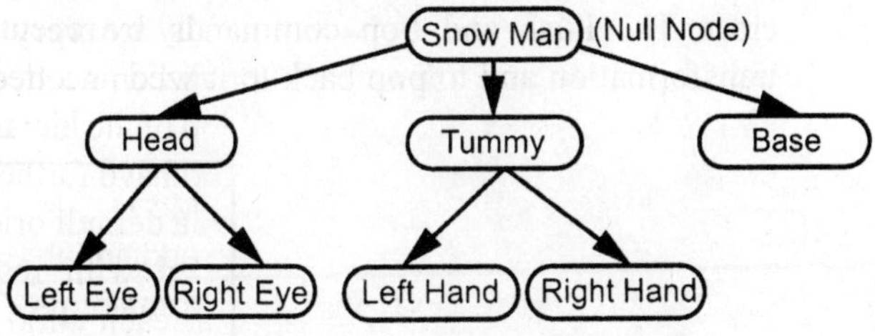

Fig.5.32: Hierarchy of the snowman model

Matrix Stacks

The concept of a stack of matrices is used for constructing hierarchical models. Matrices can be pushed on top of the stack and popped out by using the Push and Pop matrix operators. At any given time, only the topmost matrix is used to define the current transformation state. The stack being used is determined by glMatrixMode() – in this case we will just be using ModelView matrix stack. OpenGl provides two commands to deal with stacks of matrixes.

The OpenGL command

void glPushMatrix(void);

pushes all matrices in the current stack down one level. The topmost matrix (Mc in the Fig.5.33) is copied, so its contents are duplicated in both the top and second-from-the-top matrix. This is an effective way to save the current transformation state while applying further transformations at the top.

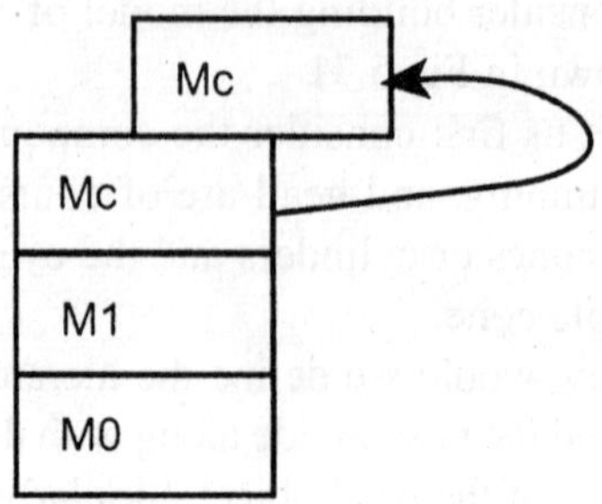

Fig.5.33: Push operator

The openGL command,

```
void glPopMatrix(void);
```

pops the top matrix off the stack, destroying the contents of the popped matrix. What was the second-from-the-top matrix becomes the top matrix. From the previous fig., we effectively pop back to the saved transformation state: matrix Mc. If the stack contains a single matrix, calling glPopMatrix() generates an error. The Push and Pop commands are very useful to 'remember the last transformation and to pop back to it when needed

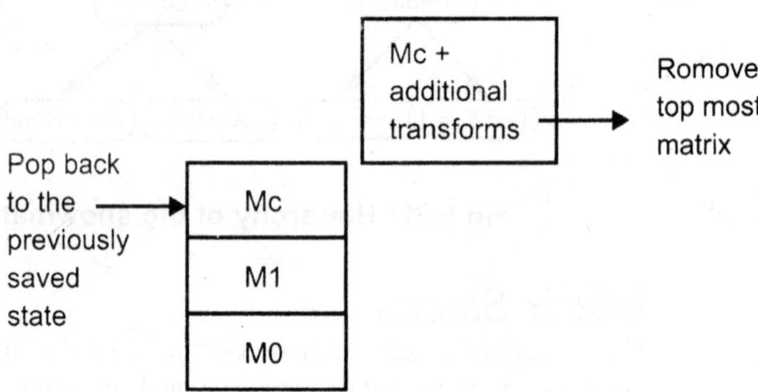

Fig.5.34: Pop operator

The Snowman Model using Hierarchy

Hierarchical models are implemented using the ModelView matrix stack.

First, the drawing routines of the building blocks are defined. The routine defines the blocks in a default position and orientation that are convenient for the modeler: the object coordinate system or *object space*. For the snowman model, the drawing routines for the building block objects—spheres, cones and disks—are already defined in the GLUT library. The sphere and disk primitives have a default position centered about the origin. The cone is based on the *x-z* plane, centered about the origin.

Within the drawing routines for each of our components, we first push the current matrix stack to remember the state. We apply the transformations (passed as parameters from the parent) for this component and finally pop back to our original state before returning from the call.

Let us see how to implement the hierarchy of Snowy using OpenGL.

We define a one dimensional array of size = 9 to define the (*T*x, *T*y, *T*z, *R*x, *R*y, *R*z, *S*x , *S*y, *S*z) transformations (also referred to as *xforms* in CG lingo) for each of the components of the snowman. The array is set to the identity transformation to begin with.

```
//Top level snowman xforms
GLfloat snomanXforms[9] - {0.,0.,0.,0.,0.,0.,1.,1.,1.};
//xform of bottom
GLfloat botXforms[9] - {0.,0.,0.,0.,0.,0.,1.,1.,1.};
// xform of tummy
GLfloat tumXforms[9] - {0.,0.,0.,0.,0.,0.,1.,1.,1.};
// xform of head
GLfloat headXforms[9] - {0.,0.,0.,0.,0.,0.,1.,1.,1.};
//Xform of eyes
GLfloat lEyeXforms[9] - {0.,0.,0.,0.,0.,0.,1.,1.,1.};
GLfloat rEyeXforms[9] - {0.,0.,0.,0.,0.,0.,1.,1.,1.};
//Xform of nose
GLfloat noseXforms[9] - {0.,0.,0.,0.,0.,0.,1.,1.,1.};
//Xform of hands
GLfloat lHandXforms[9] - {0.,0.,0.,0.,0.,0.,1.,1.,1.};
GLfloat rHandXforms[9] - {0.,0.,0.,0.,0.,0.,1.,1.,1.};
```

The drawing routine for each component accepts its corresponding xform array.

The drawing routine first saves the current transformation state by using a PushMatrix comand. It then applies its local transforms, and then pops the stack back to the original transformation state. For example, the bottom of the snowman, with a radius of 1.5, units can be defined as follows:

```
void draw_Bottom(GLfloat *botXforms){
        glPushMatrix(); // save state
        //default state of bottom
        glTranslatef(0.,1.5,0.); // translate bottom up by 1.5 units
// apply local transforms to the bottom
        glTranslatef(botXforms[0], botXforms[1], botXforms[2]);
        glRotatef(botXforms[3], 1,0,0);
        glRotatef(botXforms[4], 0,1,0);
        glRotatef(botXforms[5],0,0,1);
        glScalef(botXforms[6], botXforms[7], botXforms[8]);
        // actual drawing code
```

```
        glutSolidSphere(1.5,20,20);
        glPopMatrix(); // pop back to original state
    }
```

By default, we translate the base of the snowman up by 1.5 units, so its base is resting at the origin. Transformations to the bottom component are applied by modifying the botXforms array.

The pivot point for each component and its children is very important in a hierarchical model. Each component has a pivot point based on the location of the origin in its local object space. This point is also called the *local origin*. Local transformations for components are applied using this pivot point. In this example, the pivot point for the bottom is at the current location of the origin—the center of the sphere. If we had moved the lines of code:

```
glTranslatef(0.,1.5,0.); // translate bottom up by 1.5 units
```

to be defined just before the actual drawing code, then the local origin would be located at the base of the sphere when the transforms are applied. The pivot point for the bottom would hence be at its base.

This technique is often used to modify pivot points for components within a model. In this case, we actually wanted to re-position the sphere 1.5 units up as well. If we do not wish for the actual model to be re-positioned, we would then apply a reverse translation to move the model back to its original location. This process would modify the pivot point without affecting the position of the model. We saw this process at work in Chapter 2, Example2.4.

All children of a component are drawn within the transformation state of the parent. This ensures that the child is transformed along with its parent. Due to the Pop operation, any transformations applied to the child are not applied back to the parent. For example, the hands, which are children of the tummy, are defined within the transformation stack of the tummy. The transforms of the hands are applied only in the local routine defining the hand.

```
void draw_Tummy(GLfloat *tumXforms, GLfloat *lHandXforms, GLfloat *rHandXforms){
        glPushMatrix();
        glTranslatef(tumXforms[0], tumXforms[1], tumXforms[2]);
        glTranslatef(0.,3.9,0.);
        glRotatef(tumXforms[3], 1,0,0);
        glRotatef(tumXforms[4], 0,1,0);
        glRotatef(tumXforms[5],0,0,1);
        glScalef(tumXforms[6], tumXforms[7], tumXforms[8]);

        {
                glutSolidSphere(1,20,20);
// draw children within parent's transformation state
```

```
                draw_LHand(lHandXforms);
                draw_RHand(rHandXforms);
        }
        glPopMatrix();
}
```

The tummy has a radius of 1 unit and is transformed up by 3.9 units, so it rests on top of the bottom (!). Notice that we apply the scale and rotate transform before the translations. Doing so ensures that the pivot point for these operations is at the local origin—the center of the sphere.

The left hand (and similarly the right hand) are defined as cones. By default, they are rotated by 90 degrees and translated so that they are sticking out from the surface of the tummy. The hands can be moved by modifying the lHandsXforms array. Additionally, they will also be transformed when the tummy is! The pivot point for the scale and rotate transforms for the hand is at the local origin of the cone which is located at the center of it's base.

```
void draw_LHand(GLfloat *lHandXforms){
        glPushMatrix();
        glTranslatef(-1.,0.,0.);
        glTranslatef(lHandXforms[0], lHandXforms[1], lHandXforms[2]);
        glRotatef(lHandXforms[3], 1,0,0);
        glRotatef(lHandXforms[4], 0,1,0);
        glRotatef(lHandXforms[5],0,0,1);
        glScalef(lHandXforms[6], lHandXforms[7], lHandXforms[8]);
        //default state is pointing outward
        glRotatef(-90.,0,1,0);
        {
                glutSolidCone(0.1,1.5,5,5);
        }
        glPopMatrix();
}
```

The head and its children (eyes and nose) are similarly defined. The final snowman drawing code is as follows:

```
void draw_SnowMan(GLfloat *snomanXforms, GLfloat *botXforms, GLfloat *tumXforms, GLfloat
*headXforms, GLfloat *lXforms, GLfloat *rXforms, GLfloat *noseXforms, GLfloat *lHandXforms,
GLfloat *rHandXforms){
        glPushMatrix();
        glColor3f(1.,1.,1.);
        glTranslatef(snomanXforms[0], snomanXforms[1], snomanXforms[2]);
        glRotatef(snomanXforms[3],1,0,0);
        glRotatef(snomanXforms[4],0,1,0);
```

```
        glRotatef(snomanXforms[5],0,0,1);
        glScalef(snomanXforms[6], snomanXforms[7], snomanXforms[8]);
        {
        draw_Bottom(botXforms);
        draw_Tummy(tumXforms, lHandXforms, rHandXforms);
        draw_Head(headXforms, lXforms, rXforms, noseXforms);
        }
        glPopMatrix();
}
```

Note that the snowman object itself has no drawing component to it. Since there are no transforms before the scale and rotate calls, the pivot point is at the local origin: which happens to be the base of the snowman. This pivot is used for transforms applied to the entire snowman hierarchy (snowmanXforms).

In *Example5_13*, we rock the snowman back and forth about its base by rotating it along the *z*-axis. At the same time, we also swing its hands up and down and nod its head left to right (rotation along the *y*-axis)

```
void Display(void)
{
        GLfloat ypos, xpos;

        snomanXforms[5] + - 1*sign;
        lHandXforms[5] + - 2*sign;
        rHandXforms[5] + - -2*sign;
        headXforms[4] + - 2*sign;
        if (snomanXforms[5] - - 30)
                sign - -1;
        else if (snomanXforms[5] - - -30)
                sign - 1;

        glClear(GL_COLOR_BUFFER_BIT | GL_DEPTH_BUFFER_BIT);
        draw_SnowMan(snomanXforms, botXforms, tumXforms, headXforms, lEyeXforms,
lEyeXforms, noseXforms, lHandXforms, rHandXforms);

        glFlush();
        glutSwapBuffers();
        glutPostRedisplay();
}
```

Try playing around with the various transformations and verify that the hierarchy works as intended. The entire code can be found under *Example5_13*, in files: *Example5_13.cpp, Snowman.cpp* and *Snowman.h*. Make sure you completely understand the concept of pivot points and how they can be modified before moving on to the next section. Now, consider a slightly more complex example for modeling a hierarchical model, that of the android.

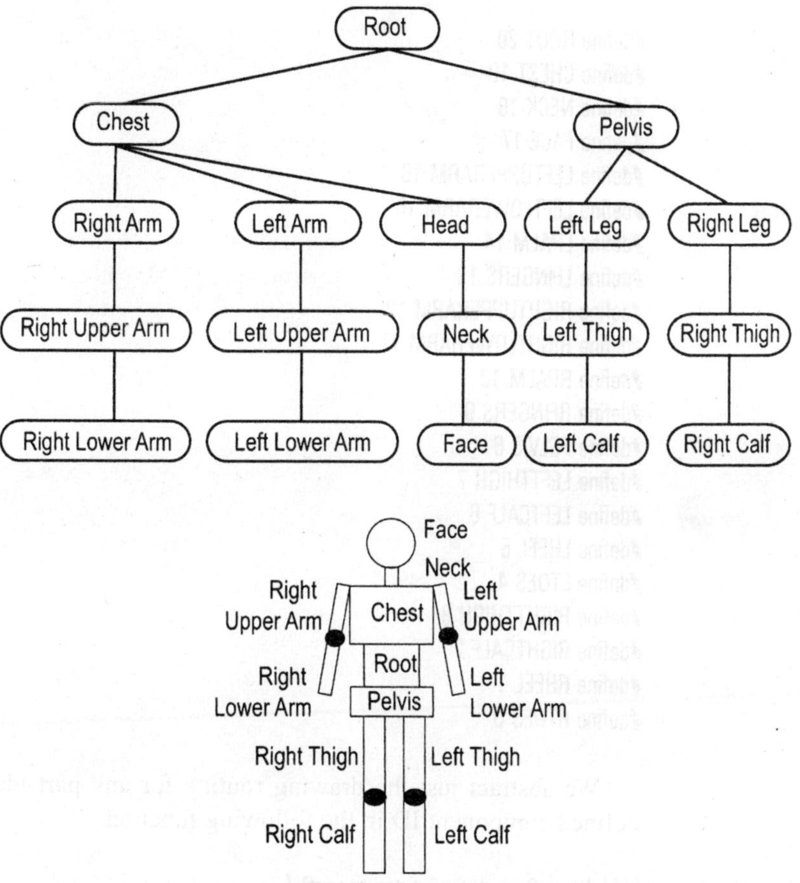

Fig.5.35: Components of the Android and its hierarchy

The entire Android is rooted at the Root component.

The lower torso is rooted at the pelvis, about which swivel its two children, the left and the right leg. The leg is composed of the thigh on which is rooted the calf.

The upper torso is rooted at the chest, which has has three children: the head, the right hand and the left hand. The head is composed of the neck upon which rotates the face. The hands can rotate beside the chest, and can be further decomposed into upper and lower arms. With this motion in mind, we can define a hierarchy for the android as shown in Fig.5.35.

By this definition, if we move the Root component, the entire android (with all its children) will be transformed. Swiveling the arms or legs will rotate the entire arm/leg, but rotating only the lower arm or leg will have no affect on its parents. Let us implement the Android model as a hierarchical model. The model has been defined in a VRML file. Each component of the model has been defined with a specific index in the coordinates data array, as shown below.

```
#define ROOT 20
#define CHEST 19
#define NECK 18
#define FACE 17
#define LEFTUPPERARM 16
#define LEFTLOWERARM 15
#define LPALM 14
#define LFINGERS 13
#define RIGHTUPPERARM 12
#define RIGHTLOWERARM 11
#define RPALM 10
#define RFINGERS 9
#define PELVIS 8
#define LEFTTHIGH 7
#define LEFTCALF 6
#define LHEEL 5
#define LTOES 4
#define RIGHTTHIGH 3
#define RIGHTCALF 2
#define RHEEL 1
#define RTOES 0
```

We abstract just the drawing routine for any part identified by the above-defined component ID in the following function:

```
void draw_Component(int component) {
int i;
for (i-0;i<noofpoly[component]*3;i-i+3) {glBegin(GL_TRIANGLES);
glVertex3fv(&(coords[component][3*indices[component][i]]));
glVertex3fv(&(coords[component][3*indices[component][i+1]]));
glVertex3fv(&(coords[component][3*indices[component][i+2]]));
glEnd();}
}
```

Let us start defining the actual components. For this model, we do not define an array of transforms; each component has only specific transforms that can be applied to it. The right leg is composed of the calf as its leaf node (we do not maintain the toes and heel as further nodes in this hierarchy). The calf can only rotate about the x-axis (swing back and forth). Since the calf is not defined with its base at the origin, we do need to translate it to the origin before rotating it (and translate it back when we are done). This will ensure the correct pivot point being used for rotation. The component RightCalf is defined as follows:

```
void draw_RightCalf(GLfloat angx) {
// save the current transformation state
  glPushMatrix();
// translate the component to the origin before rotation: pivot point
  glTranslatef(-.14,-1.59,0.1);
  glRotatef(angx, 1.,0.,0.);
//  translate the component back to the original position
  glTranslatef(.14,1.59,-0.1);
// draw the actual geometry in this transformed space
    draw_Component(RIGHTCALF);
    draw_Component(RHEEL);
    draw_Component(RTOES);
// reset the transformation stack
  glPopMatrix();}
```

The thigh can rotate about both the *x*- and *z*- axis for a forward and sideways swing respectively. The function to draw this component is defined below. In addition to its own transforms, we also pass in the rotation information for its child node.

```
void draw_RightThigh(GLfloat angx,GLfloat angz,GLfloat calfang) {
  glPushMatrix();
  glTranslatef(-.2,-.7,0.);
  glRotatef(angx, 1.,0.,0.);
  glRotatef(angz, 0.,0.,1.);
  glTranslatef(.2,.7,0.);
// Draw the component and all its children in this transformation space
  draw_RightCalf(calfang);
  draw_Component(RIGHTTHIGH);
  glPopMatrix();
}
```

The top-level component, RightLeg, does not have any geometry associated with it and is defined as follows:

```
void draw_RightLeg(GLfloat *rightlegang) {
 draw_RightThigh(rightlegang[0], rightlegang[1],rightlegang[2]);
}
```

The left leg and the arms are similarly defined with the appropriate hierarchy. The head consists of the neck and the face. We define the face to be a child of the neck. We allow it to swivel about the *y*-axis, causing it to revolve from side to side. We do not allow the neck any freedom to transform.

```
void draw_Face(GLfloat angy) {
 glPushMatrix();
  glRotatef(angy, 0.,1.,0.);
  draw_Component(FACE);
 glPopMatrix();
}

void draw_Neck(GLfloat headang) {
 draw_Face(headang);
 draw_Component(NECK);
}

void draw_Head(GLfloat headang){
  draw_Neck(headang);
}
```

Notice that we do not define any limits to the amount one can swivel the components. For realistic motion you should clamp the rotations so you don't end up twisting the body in weird ways!

And finally the android itself can be defined as follows:

```
void draw_Android(GLfloat *T, GLfloat *R,
      GLfloat headang,
      GLfloat *leftarmang, GLfloat *leftlegang,
      GLfloat *rightarmang, GLfloat *rightlegang)
{
        glPushMatrix();
        // transform the entire android
        glTranslatef(T[0], T[1], T[2]);
        glRotatef(R[0],1.,0.,0.);
        glRotatef(R[1],0.,1.,0.);
        glRotatef(R[2],0.,0.,1.);
        // draw the components and their children
        draw_Component(ROOT);
        {
                draw_Component(CHEST);
                draw_RightArm(rightarmang);
                draw_LeftArm(leftarmang);
                draw_Head(headang);
        }
        {
                draw_Component(PELVIS);
                draw_LeftLeg(leftlegang);
```

```
                        draw_RightLeg(rightlegang);
            }
            glPopMatrix();
}
```

In *Example5_14*, we make Android march by continuously changing the rotation on the arms and legs.

```
void Display(void)
{

  if (cycle%2 ==0) {
    rightarmang[0]-=0.3; leftarmang[0]+=0.3;
    rightarmang[2]-=0.4; leftarmang[2]+=0.4;
    rightlegang[0]-=0.6; leftlegang[0]+=0.6;
    rightlegang[2]+=1.2; leftlegang[2]-=1.2;
  } else {
    rightarmang[0]+=0.3; leftarmang[0]-=0.3;
    rightarmang[2]+=0.4; leftarmang[2]-=0.4;
    rightlegang[0]+=0.6; leftlegang[0]-=0.6;
    rightlegang[2]-=1.2; leftlegang[2]+=1.2;
  }
  pos++;
  if (pos >=100) {
    pos = 0;
    cycle++;
  }

  glClear (GL_COLOR_BUFFER_BIT);
  glColor3f (1.0, 1.0, 1.0);
  glLoadIdentity();
  draw_Android(T, R, headang, leftarmang, leftlegang, rightarmang, rightlegang);
  glFlush();
  glutSwapBuffers();
  glutPostRedisplay();
}
```

You can experiment with different kinds of transformations on the Android to verify that the motion is consistent with the model hierarchy we designed. Ideally, we would like to move the Android based on some user input. We leave this part of the code as an exercise for the reader. The entire code can be found in *Example5_14/Example5_14.cpp.* You will have to compile and link this program with the provided files: *vrml.cpp* and *vrml.h.*

Summary

In this chapter, we have learned the basics of 3D modeling. The polygon shape was used to define primitive as well as complex shapes. We have learned how principles of transformations can be used to define hierarchical models. By developing 3D models, defining a viewing position, and a projection, we can define a 3D world inside our computer. The camera just needs to be clicked to photograph the world and display it onto the screen. In the next chapter we shall see how to render this 3D world to generate realistic images.

Chapter 6

Rendering: Shading and Lighting

In the last chapter, we saw how to model objects and represent them as wire-frame models. Wire frame models depict the outer hull of the object but do not convey a realistic image of the model. The next step in our quest for visual realism is to paint our models so we can render realistic images. This process is called *rendering*.

In this chapter you will learn all about the components of rendering:

- Hidden surface removal
- The CG reflectance model
- Surface materials (including texture mapping)
- Lights
- Shading algorithms

6.1 What is Rendering?

Rendering 3D images is a multi-step process. First, we need to identify which component surfaces (usually polygons) of the model are viewable from the current viewpoint. This process involves back-face culling as well as identifying surfaces obstructed by surfaces in front of them. Once visible surfaces have been identified, we can simply assign a color to them and paint them. We saw this approach to some extent in Chapter 1.

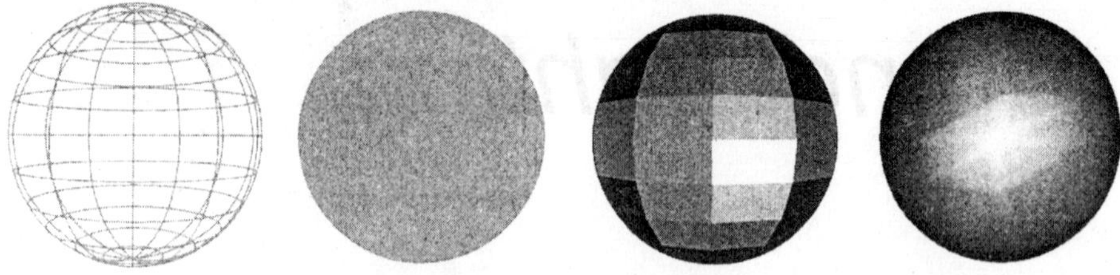

Fig. 6.1: A sphere rendered with a) a single color, b) each polygon rendered with a single color, c) shaded smoothly

In most cases, however, we do not want surfaces to be colored with just one color. Surfaces should appear shaded based on the amount of light that they receive, as is the case in real life. To be able to simulate this effect, we need to define material properties of the surface: not only its color but also how it responds to light. This process is called *shading*. We need to define light sources to light up the scene, which enable us to view it—a process called *lighting*. Once the shading and lighting of a scene has been established, shading algorithms are then used to finally render the images.

Rendering is a complex process and can be very time consuming. For example, a typical image in the film *Toy Story* took anywhere between an hour and 72 hours to render on a workstation! Different shading algorithms use different methods and tricks to simulate the behavior of light and surfaces. Obviously, which algorithm you use is based on your precise needs. An interactive game needs shaders whose output may not be very sophisticated but can be rendered quickly, whereas a blockbuster movie production can afford to use complex shaders that can take days to render images but produces spectacular results. We shall employ some simple shading algorithms in this book and shall introduce you to some more complex ones in Chapter 7.

Let us first start with a more in-depth discussion of hidden surface removal.

6.2 Hidden Surface Removal

In the previous chapter, we looked at wire frame representations of models. The models we constructed were "see through"—we could see through the polygons. We learned how to eliminate the back-facing polygonal surfaces of our model using a technique called back-face culling. But what about those surfaces that are obscured by other objects in front of them as shown in Fig..?

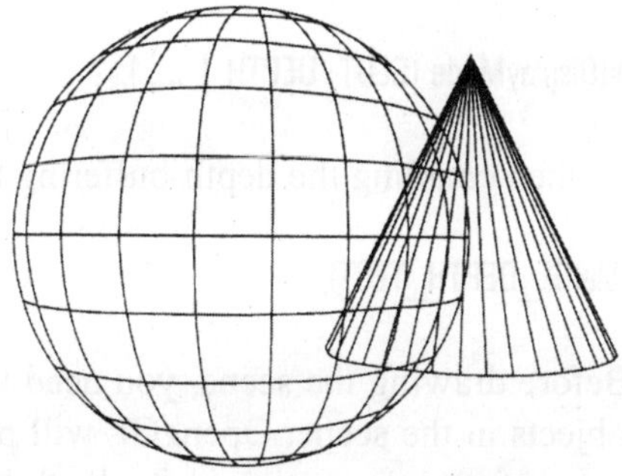

Fig. 6.2: A sphere obstructing a cone: back-face culling is on

In a wire frame mode, it probably is still okay to see through the first object. But when we draw our object as a solid surface, we want to see the object that is in front. Compile and run *Example6_1/Example6_1.cpp*.

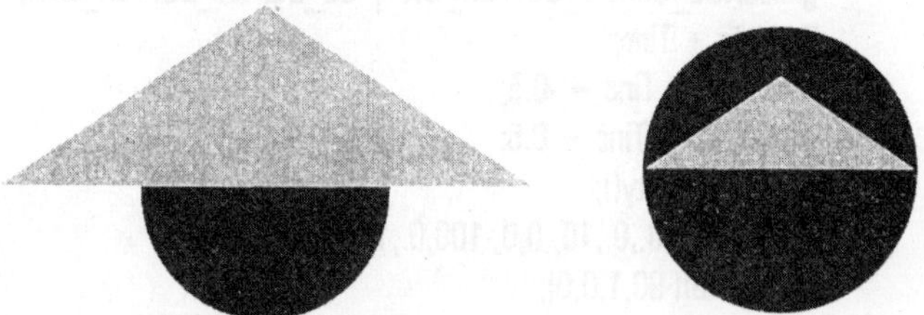

Fig. 6.3: A cone a. in front of the sphere, b. behind the sphere

This example shows a solid cone and a sphere. The cone translates back and forth along the *z*-axis. Surprise, surprise- even when the cone is behind the sphere, we still see it being drawn in front! We want not only the back surfaces of the sphere and cone to be hidden from view, but also the portion of the cone being obstructed from view by the sphere. The process to remove these surfaces is called *hidden surface removal*. Back-face culling is part of this removal process.

Z-Buffering

One of the simplest techniques to accomplish hidden surface removal is known as *z-buffering* or *depth buffering*. In this process, a buffer in memory called the z-buffer is used to keep track of the surface that is closest to the eye for any given pixel. The closest object finally gets drawn onto the screen. Note that for solid closed objects, this process will automatically remove back-facing surfaces also. OpenGL provides z-buffering by way of a depth buffer. To use it, you need to first enable depth buffering. You can do this by initializing the display mode to use the depth buffer

```
glutInitDisplayMode (GLUT_DEPTH | .... );
```

—and then enabling the depth-buffering test

```
glEnable(GL_DEPTH_TEST);
```

Before drawing the scene, you need to clear the depth buffer and then draw the objects in the scene. OpengGL will perform all the calculations to store the closest surface information in its depth buffer and draw the scene accordingly. Shown below is a snippet of code to draw a sphere and a cone. The cone is animated to periodically move in front of and then behind ahead the sphere. Depth buffering ensures that obstructed parts do not get drawn onto the screen. The entire code can be found under *Example6_1/Example6_1.cpp* (you will need to remove the commented lines to enable the depth test)

```
glClear(GL_COLOR_BUFFER_BIT | GL_DEPTH_BUFFER_BIT);
Tz - Tz + Tinc;
  if (Tz > 5) Tinc - -0.5;
  if (Tz < -5) Tinc - 0.5;
  glLoadIdentity();
  gluLookAt(0.,0.,10, 0,0,-100,0.,1.,0.);
  glRotatef(-90,1,0,0);
  {
    glColor3f(1.,0.,0.);
    glutSolidSphere(2., 8,8);
  }
  {
    glColor3f(0.,1.,0.);
    glTranslatef(0.,Tz,0);
    glutSolidCone(3.,2.,8,8);
}
```

When you make a call to glColor, all surfaces drawn after the call will use the same color. If you see how the cone and sphere are rendered, you will be very disappointed. They look so 2D! We would like our surfaces to be shaded to give a sense of depth.

Surface removal is but one factor in generating a realistic image. Our next task is to shade the visible surfaces so they appear 3D based on the lights illuminating them, and their surface material properties. Since we want to simulate the real world, CG simulation of lighting and shading follows a close analogy to how lights and surfaces behave in the real world, and how humans actually see. Let us see how light works in the real world and then draw upon this analogy in our discussions of CG based lights and materials.

6.3 Light: Reflectance Model

Imagine entering a dark room with no lights in it. If the dark room is perfectly insulated from any kind of light, you will not be able to see anything in the room even though there are objects in it.

Imagine now that we switch on a bulb in the room. Immediately, we shall

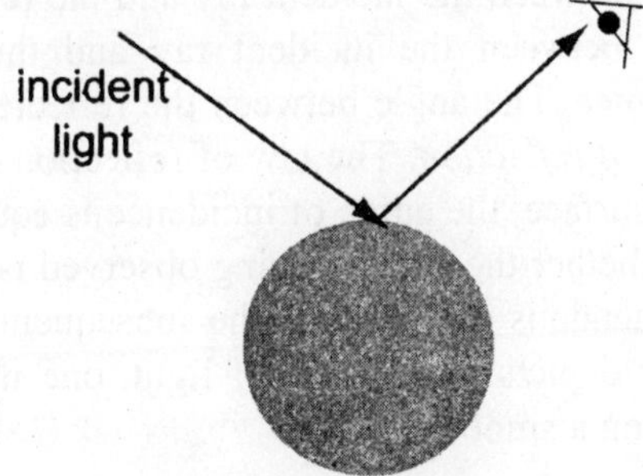

Fig. 6.4: Light rays entering the eye enable sight

start seeing the objects in the room. How brightly we see them will depend on the wattage and color of the bulb as well as on the material that the objects are made of. The same object may appear reddish under a red light or eerily blue under a blue light. We are able to see because light bounces off (reflects) from objects and eventually reaches our eyes. Once light reaches our eyes, signals are sent to our brain, and our brain deciphers the information in order to detect the appearance and location of the objects we are seeing. The light sources directly or indirectly define the incident light rays. .The surface properties of the objects in the room, also called the *surface material*, determine how the incoming light is reflected. Let us explore the way light is reflected in further detail.

Law of Reflection

Light is usually modeled as light rays (which can be thought of as vectors!). Many rays traveling together are referred to as a beam of light. Light rays behave in a very predictable manner. If a ray of light could be observed approaching and reflecting off a flat mirror, then the behavior of the light as it reflects would follow the the law of reflection.As shown in Fig.6.5, a ray of light, *I*, approaches

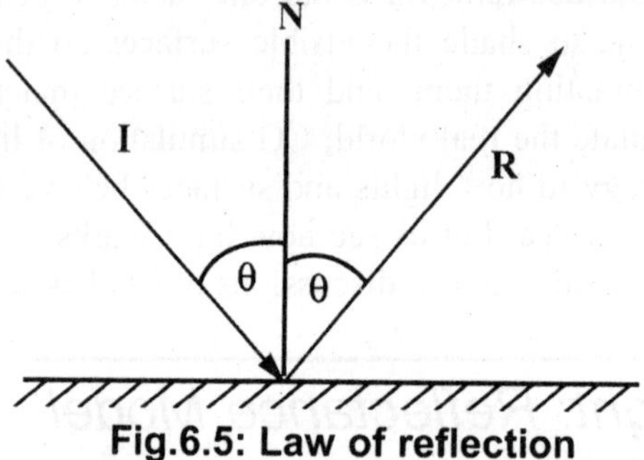

Fig.6.5: Law of reflection

the mirror and is called the *incident ray*. The ray of light, *R*, that leaves the mirror, is known as the *reflected ray*. At the point of incidence where the ray strikes the mirror, we can define a normal, *N*, to the surface of the mirror. The normal is perpendicular to the surface at the point of incidence and divides the angle between the incident ray and the reflected ray into two equal angles. The angle between the incident ray and the normal is known as the *angle of incidence*. The angle between the reflected ray and the normal is known as the *angle of reflection*. The law of reflection states that when a ray of light reflects off a surface, the angle of incidence is equal to the angle of reflection.

Whether the surface being observed is microscopically rough or smooth has a tremendous impact upon the subsequent reflection of a beam of light. Fig. 6.6 below depicts two beams of light, one incident upon a rough surface and the other on a smooth surface.

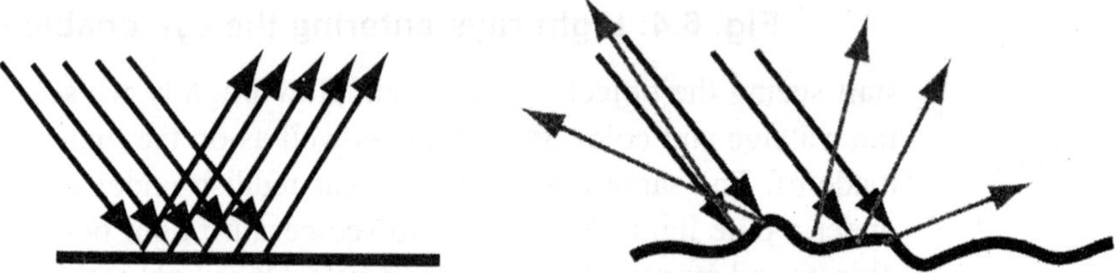

Fig.6.6: Reflection off smooth and rough surfaces

For both cases, each individual ray follows the law of reflection. For smooth materials, such as mirrors or a calm body of water, all the rays reflect and remain concentrated in the same bundle upon leaving thc surface, causing the surface to have a shiny look. This type of reflection is known as *specular reflection*.

For rough materials, such as clothing or an asphalt roadway, the roughness of the material means that each individual ray meets a surface with a different orientation. The normal line at the point of incidence is different for different rays. Subsequently, when the individual rays reflect according to the law of reflection, they scatter in different directions. The result is that the rays of light incident to the surface are diffused upon reflection. This type of reflection is called *diffuse reflection*. In practice, both types of reflection can be observed on the same surface.

When light strikes an object, not only does a portion of it gets reflected, as discussed above, but it can also be absorbed or transmitted through. Objects have a tendency to selectively absorb, reflect, or transmit light of certain colors. So, one object might reflect green light while absorbing all other frequencies of visible light, causing us to see it as green. Another object might selectively transmit blue light while absorbing all other frequencies of visible light, causing it to look blue and translucent.

Light being bounced around tends to *decay* over time and distance due to atmospheric absorption. This is why far-away objects appear dimmer than objects that are closer to us. This phenomenon is referred to as *attenuation*.

The color and intensity of the reflected light that eventually reaches the eye is the color that the surface is perceived to have.

6.4 CG: Reflectance Model

In CG we try to emulate the above-defined reflectance model to approximate the way lights and surface materials. Mathematical equations that approximate this behavior govern the final color of each pixel in the image. The most popular CG reflectance model is the one proposed by *Phong*.

Phong Reflectance Model

The Phong reflectance model is also the model employed by OpenGL. In this reflectance model, we break reflectance into four components:

1. **Ambient Reflectance:** This is a hack introduced in CG. It is used to avoid CG scenes from going completely black. Ambient light has no direction; it impinges equally on all surfaces from all directions and is reflected by a constant multiple, resulting in flat-looking surfaces.
2. **Diffuse Reflectance:** Surfaces with diffuse reflectance scatter light equally in all directions, but the intensity of the reflection varies based on the angle of incidence.
3. **Specular Reflectance:** Specular reflections can be observed on any shiny surface, causing a dull grey or white highlight on the surface.
4. **Emission:** The emission component of a surface is again a hack. It is used to simulate luminous objects that seem to glow from their own light, such as a light bulb.

Within our CG scene, we define surface materials with reflection coefficients for each of the four components. The reflection coefficients are defined as RGB triplets, and have a value between 0 and 1 – 1 being maximum reflection. Many people refer to these reflectance coefficients as the color of the material, since it will be the color of the light it will reflect. This material is then assigned to models in the scene.

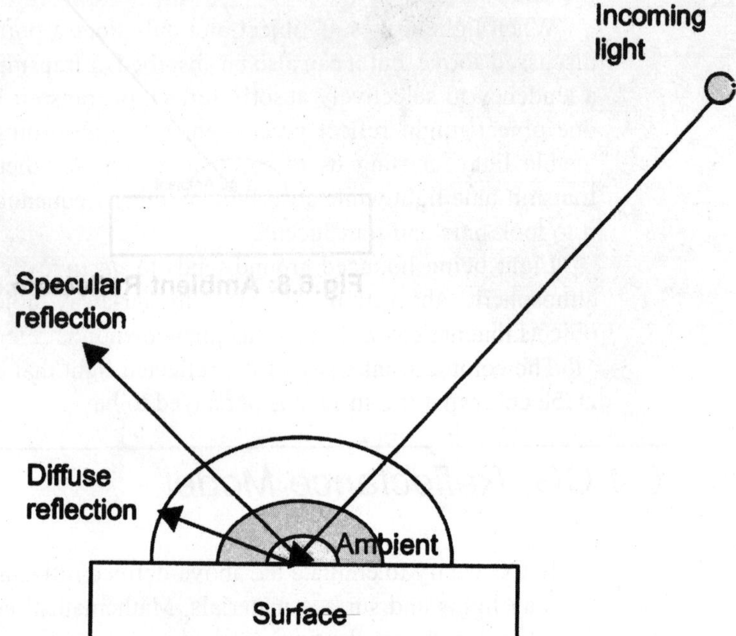

Fig.6.7: Four components of Phong reflectance model

CG lights are defined to illuminate the scene. These lights are again modeled with RGB colors for the ambient, diffuse, and specular components. The red, green and blue components have a value between 0 and 1. Some software packages further define an intensity to the light: the intensity is simply a scale factor for the color of the light: the actual color of the light is the product of its color and its intensity. The OpenGL model does not use intensity for light sources.

When incoming light strikes an object, each of these four components is calculated to determine its individual contribution to the reflected light. All four components then added together to attain the final color of the outgoing light

Let us look into each of these components and how they are calculated.

Ambient Reflectance

Ambient reflectance is actually just an approximation to the fact that there is never a perfect dark room in this world. Some amount of light always seeps in, enabling vision. This kind of illumination enables us to see faint traces of objects with flat shades of surface colors and is referred to as *ambient lighting.*

When light with an ambient component (ambient light) strikes a surface with ambient reflectance, it is scattered equally in all directions with a constant multiple, causing the surface to look flat as shown in Fig.6.9.

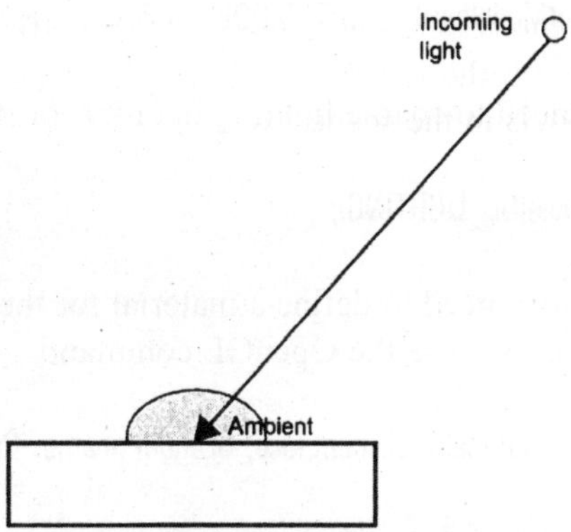

Fig.6.8: Ambient Reflectance

Mathematically, if the incident light has a an ambient color of (R_i,G_i, B_i) and the object it hits has a material with ambient reflectance of (R_a, G_a, B_a) then the reflected light due to ambient reflection will have a color defined as (R_{ra}, G_{ra}, B_{ra}) = (R_i*R_a, G_i*G_a, B_i*B_a).

Fig.6.9: Ambient lighting on a sphere

Let us see how to render a sphere using ambient lighting. In openGL, we can specify global lighting. Global lighting is not associated with any light source, but is prevalent uniformly in the 3D world being defined. The command to set up a global ambient light is

```
glLightModelfv(GL_LIGHT_MODEL_AMBIENT, lmodel_ambient);
```

where lmodel_ambient is a vector defining the ambient RGBA intensity. A is the alpha component of the color. We will not be using it in this chapter, so always set this value to 1. By default, the value of the ambient color is

```
lmodel_ambient[] - {0.2, 0.2, 0.2, 1.0};
```

In *Example6_2*, we use global ambience to light up a sphere. The init function defines the lighting model as ambient and sets the ambient color to a dull white, as shown below.

```
GLfloat light_ambient[] - { 0.5, 0.5, 0.5, 1.0};
```

```
glLightModelfv(GL_LIGHT_MODEL_AMBIENT, light_ambient);
```

We need to enable lighting in order for it to affect the rendering as shown below

```
glEnable(GL_LIGHTING);
```

We also need to define a material for the sphere. To set material properties of an object, we use the OpenGL command

```
void glMaterialfv(GLenum face, GLenum pname, TYPE* param);
```

This command specifies a current material property for use in lighting calculations.

face can be GL_FRONT, GL_BACK, or GL_FRONT_AND_BACK to indicate which face of the object the material should be applied to.

pname specifies the particular material property being (such as GL_AMBIENT for ambient reflectance)

param is the desired value(s) for the property specified.

In our example, we set the ambient reflectance of the material such that it reflects all the red component of incident light and absorbs all of green and blue by defining the material as

```
GLfloat mat_ambient[] - { 1, 0.0, 0.0, 1.0};
glMaterialfv(GL_FRONT, GL_AMBIENT, mat_ambient);
```

Since it reflects only the red component of light, the material will look red. Indeed, you can see the analogy between the color of the material and its reflectance coefficients. Surfaces are rendered using the most recently defined material setting. Since we will only be setting one material for this example, we can set it in the init() function itself. The display code is as follows:

```
void Display(void)
{
        glClear (GL_COLOR_BUFFER_BIT | GL_DEPTH_BUFFER_BIT );
        // material as already been set in the init function
        glutSolidSphere(2., 40,32);
        glFlush();
}
```

Remember to always turn on depth buffering when lighting 3D scenes. This will ensure you are always viewing the surfaces that are closest to the camera! Run the program. Notice how the sphere appears as a dull flat colored red ball.

The sphere looks flat because it only has ambient reflection defined. The entire code can be found under *Example6_2/Example6_2.cpp*.

Diffuse Reflectance

Surfaces with diffuse reflectance scatter light equally in all directions. The amount of light reflected is directly proportional to the angle of incidence of the incoming beam of light. Diffuse reflectance can be observed on dull objects such as cloth or paper.

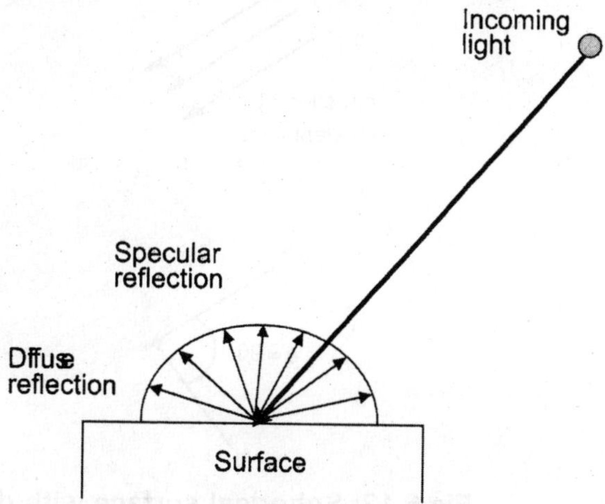

Fig.6.10: Diffuse reflectance

The exact math to calculate diffuse reflection was proposed by Lambert and so this reflectance is often referred to as *Lambert reflectance*. If we assume the light to have a diffuse color of (R_{id}, G_{id}, B_{id}) and the surface to have a diffuse reflection coefficient of (R_d, G_d, B_d). then the diffuse color component of the reflected light can be calculated as

$$(R_{rd}, G_{rd}, B_{rd}) = (R_{id}R_d * \cos(\theta), G_{id}G_d\cos(\theta), B_{id}B_d\cos(\theta))$$

where θ is the angle between the incident ray (direction of light source) and the normal to the surface at the point of incidence. So if θ is 0 (the light hits the

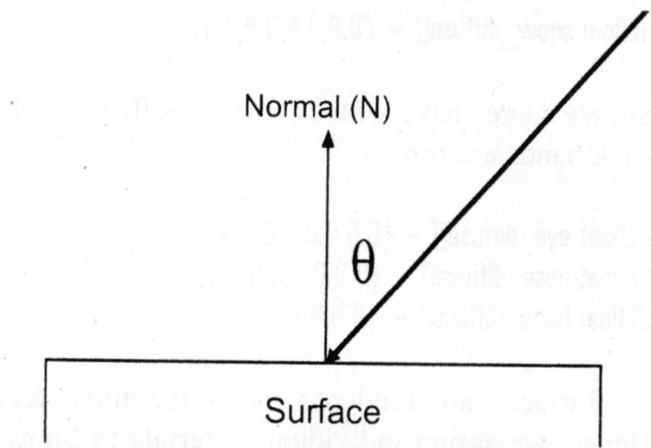

Fig.6.11: θ used in the diffuse reflection calculation

surface straight on) then the diffuse reflection is the brightest, but light incident on the surface at more than 90 degrees (cos(>90) <= 0) causes no light to be reflected, and the corresponding area appears black. The diffuse color component of a light essentially models the color of the light as we perceive it.

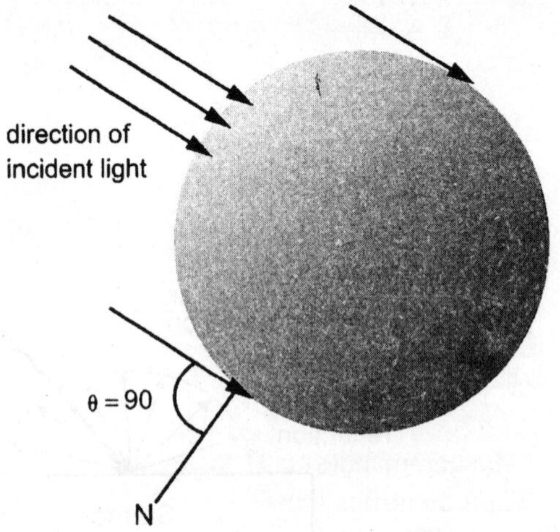

Fig.6.12: Spherical surface with diffuse reflectance

Let us apply ambient and diffuse reflectance to our famous snowman from Chapter 5. In *Example6_3*, we define all parts of Snowy to have the same ambient reflectance. The ambient color is slightly blue and is defined in the top-level function, *draw_Snowman* as

```
GLfloat mat_ambient[] = { 0.1, 0.1, 0.2, 1.0 };
glMaterialfv(GL_FRONT, GL_AMBIENT, mat_ambient);
```

The diffuse components are different for each part. The snowballs have a whitish diffuse component:

```
GLfloat snow_diffuse[] = {0.8,0.8,0.8,1.};
```

Snowy's eyes have a black diffuse reflectance his carrot nose is orange and his stick hands are brown

```
GLfloat eye_diffuse[] = {0.0,0.0,0.0,1.};
GLfloat nose_diffuse[] = {0.9,0.5,0.0,1.};
GLfloat hand_diffuse[] = {0.5,0.3,0.1,1.};
```

Surfaces are rendered using the most recently defined material setting. Hence, we assign individual materials to Snowy's parts in their local drawing routine. For example, the draw_Bottom function is now defined as

```
void draw_Bottom(GLfloat *botXforms){
        glMaterialfv(GL_FRONT, GL_DIFFUSE, snow_diffuse);
```

and the draw_Nose function is redefined as

```
void draw_Nose(GLfloat *noseXforms){
        glMaterialfv(GL_FRONT, GL_DIFFUSE, nose_diffuse); etc.
```

We now need to define a light with a specific direction to see the effect of diffuse reflectance. The OpenGL command to specify a light source and its properties is either glLightf for nonvector proprties or glLightfv for vector properties:

```
void glLightf(GLenum light, GLenum pname, TYPE param);
void glLightfv(GLenum light, GLenum pname, TYPE *param);
```

The command(s) take three arguments.

light identifies the light source and can have one of 8 values: GL_LIGHT0, GL_LIGHT1,..., GL_LIGHT7.

pname identifies the characteristic of the light being defined. The characteristics of lights include the component intensity, position and attenuation of the light source, as well as certain spotlight settings.

param indicates the value that the characteristic is set to: it's a pointer to a group of values if the vector version is used, or the value itself if the nonvector version is used.

We define one light source, GL_LIGHT0 to light up our snowman. We define this light source to have an ambient as well as diffuse color of dull white

```
GLfloat light_ambient[] = { 0.5, 0.5, 0.5, 1.0 };
GLfloat light_diffuse[] = { 0.8, 0.8, 0.8, 1.0 };
glLightfv(GL_LIGHT0, GL_AMBIENT, light_ambient);
glLightfv(GL_LIGHT0, GL_DIFFUSE, light_diffuse);
```

We need to enable lighting as well as the light source chosen:

```
glEnable(GL_LIGHTING);
glEnable(GL_LIGHT0);
```

To set the direction of our light, a vector of four values (*x, y, z, W*) is supplied with a GL_POSITION parameter to the glLightfv function. If the last value, *W*, is zero, the corresponding light source is a directional one and the (*x, y, z*) values describe its direction. We define the direction of our light source to be coming along the *z*-axis:

```
GLfloat light_position[] = {0., 0.0, 1, 0.0};
```

```
glLightfv(GL_LIGHT0, GL_POSITION, light_position);
```

Note that the position and direction of lights are transformed by the modelview matrix! Let us make our graphics more interesting by having the light rotate about the sphere. To do this, we define the position of the light source in the Display function (within the Model space), and constantly spin it around.

```
glPushMatrix (); // save current transform
glRotatef ( spin, 0.0, 1.0, 0.0); //rotation applied to light
glLightfv (GL_LIGHT0, GL_POSITION, light_position);
glPopMatrix (); // pop back to saved transform
draw_SnowMan(..)
```

The spin variable increments at every tick. The entire code can be found under *Example6_3*, in files *Example6_3 cpp*, *Snowman.cpp* and *Snowman.h*.
Voila! You will see a snowman that is lit up by a rotating light source

Specular Reflectance

Specular reflection produces shiny highlights often seen on shiny objects like glass or metal. Specular reflection, unlike ambient and diffuse reflection depends on both the direction of the incident light as well as the direction that the surface is being viewed from. The reflection is brightest when the viewing direction is parallel to the reflected light.

The equations to define specular reflectance have been proposed by many mathematicians. Each formula simulates the specular reflectance slightly differently. However, in general, specular reflection depends on three components:

1.surface orientation, *N*

2.direction of light source, *I*

3. viewing direction, *V*

Assuming a function specular that depends on the above three factors, the specular reflection color can be defined as

$(R_{rs}, G_{rs}, B_{rs}) = (R_{is}*R_s*specular(), G_{is}G_s*specular(), B_{is}B_s*specular())$

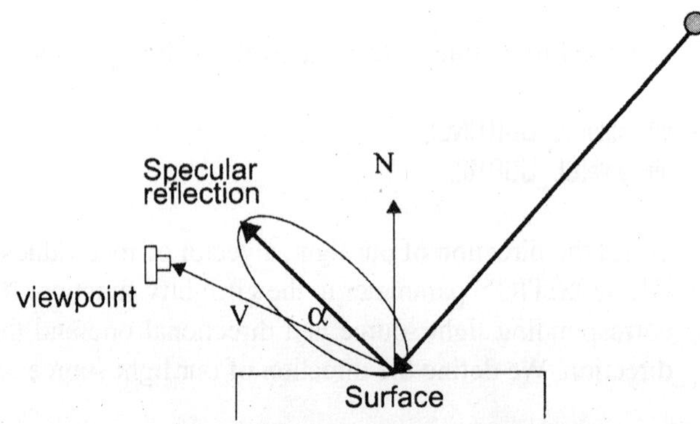

Fig.6.13: Specular reflectance

Phong proposed that this function can be approximated to be $\cos(\alpha)^n$, where α is as shown in Fig.6.13, and n is the specular exponent that determines the size of the highlight. Other equations have been proposed by Blinn and others to model specular reflections as well. Each equation output shiny highlights, but the softness of the highlight differs.

Fig.6.14: Specular reflection

The GL_SPECULAR parameter affects the color of the specular highlight.

Let us see how to implement a specular lighting scheme in openGL. Typically, a real-world object such as a glass bottle has a specular highlight that is whitish. Therefore, if you want to create a realistic effect, set the GL_SPECULAR parameter of the light to a dull white. By default, GL_SPECULAR is defined to be (1.0, 1.0, 1.0, 1.0) for GL_LIGHT0 and (0.0, 0.0, 0.0, 0.0) for any other light.

We can modify *Example6_3*, by adding a specular component to our GL_LIGHT0 as follows:

```
GLfloat light_specular[] = {0.8, 0.8, 0.8, 1.0};
glLightfv(GL_LIGHT0, GL_SPECULAR, light_specular);
```

The material for all Snowy's components have a common specular reflective component, defined as

```
GLfloat mat_specular[] = {1.0, 1.0, 1.0, 1.0};
glMaterialfv(GL_FRONT, GL_SPECULAR, mat_specular);
```

OpenGL allows you to control the size and brightness of the highlight using the parameter: GL_SHININESS. You can assign a number in the range of [0.0, 128.0] to GL_SHININESS—the higher this value, the smaller and brighter (more focused) the highlight. We define our material to have a shininess defined as

```
GLfloat low_shininess[] = {20};
glMaterialfv(GL_FRONT, GL_SHININESS, low_shininess);
```

When you run the program, you will see a shiny snowman as shown in Fig.6.15

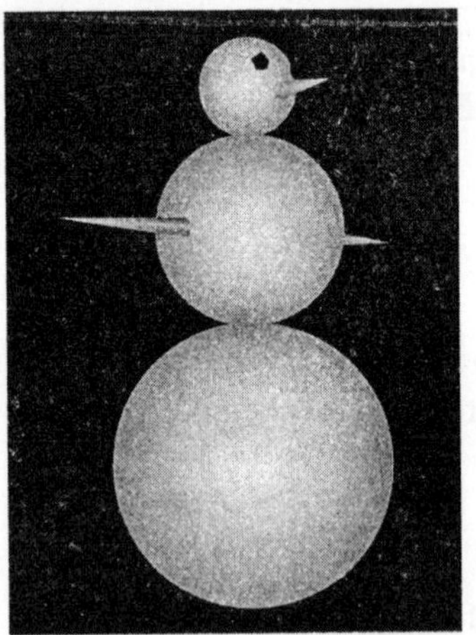

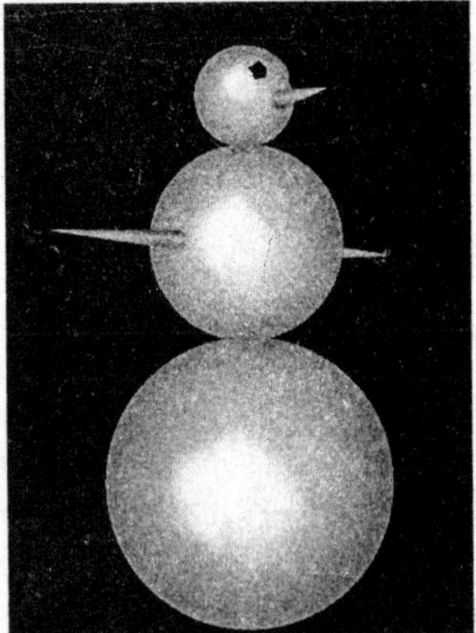

Fig.6.15: Snowy with diffuse and specular reflection

Emission

The emissive color of a surface adds intensity to the object, but is unaffected by any light sources.

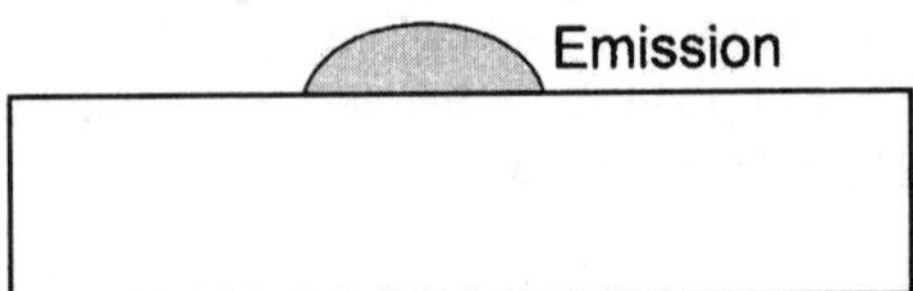

Fig.6.16: Emissive component of reflectance

Since most real-world objects (except lights) don't emit light, you'll probably use this feature mostly to simulate lamps and other light sources in a scene. In *Example6_3*, we can add a red emissive color to Snowy's nose (he has a cold, poor fellow) as follows:

```
GLfloat mat_emission[] = {0.5, 0.0, 0.0, 0.0};
void draw_Nose(GLfloat *noseXforms){
        glMaterialfv(GL_FRONT, GL_DIFFUSE, nose_diffuse);
        glMaterialfv(GL_FRONT, GL_EMISSION, mat_emission);
```

When you run the program, you will see a glow in the dark nose. Scenes can be pretty boring with just one kind of light illuminating them. In real, we see lights of all kinds—distant lights such as the sun, spotlights such as light emanating from a light bulb or those used in theater. All these lights help to add to the three-dimensionality of our world. Let us explore what more OpenGL can provide to us in terms of lights.

OpenGL Lights

OpenGL can assign many more attributes to lights than just their color. Lights can be defined with a position, a direction, and even a cone of illumination etc. The most commonly used types of lights in OpenGL are

- Distant (directional) light
- Point light
- Spotlight

Distant Light

The best example of a distant light is the sun. As viewed from any one position on the earth, its rays are essentially parallel, and the position makes no practical difference on the intensity of the light. In Fig.6.17 a distant light is shown lighting up a sphere. The distant light is represented by directed lines to indicate which direction the light is coming from. Notice how the side of the sphere facing away from the light is completely black.

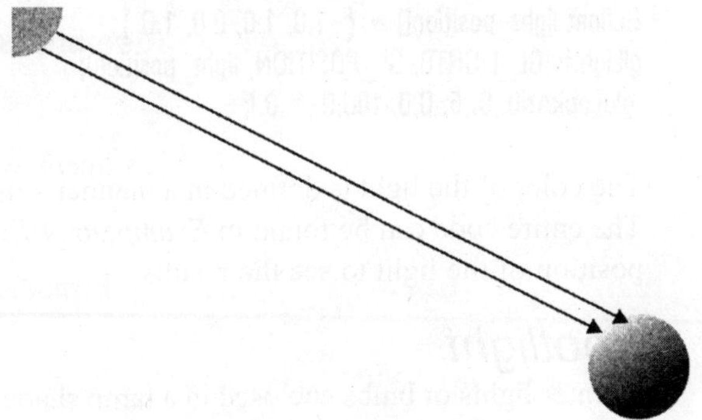

Fig.6.17: Distant Light

The light rays from a distant light flow uniformly in space from one direction. As a result, surfaces with the same orientation receive the same amount of light independent of location. Surfaces with different orientations are illuminated differently. Those facing toward the light source appear brightest, while those facing away are completely un-illuminated by this light source within the scene. In OpenGL, distant light sources are modeled with a color and a direction. We saw an example of how to use a distant light when we were illuminating our snowman in *Example6_3*.

Point Light

A bulb is a good example of a point light. A point light source distributes light through space from a single point. It distributes beams of light evenly in all directions. The intensity of the light is usually defined to fall off with the square of the distance from the light to the surface. You could, however, define the intensity to have no fall off at all. A point light needs a position to be completely defined, and is often referred to as positional light.

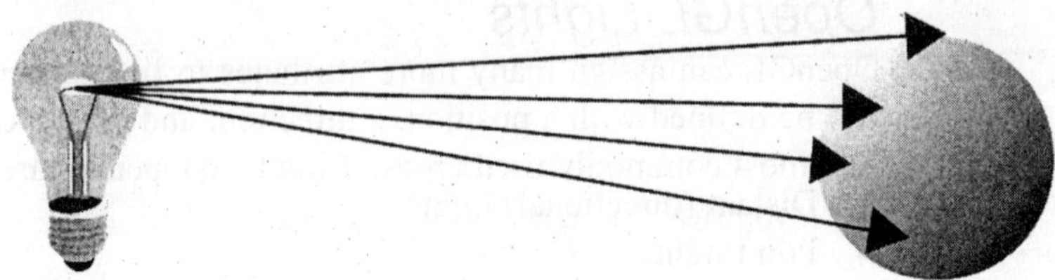

Fig.6.18: Point Light

In *Example 6_4*, we define a light source located at (-1,0,0) to light up a red sphere (which is located at the origin). To define the position of the light, we set the vector of *(x,y,z,W)* values for the GL_POSITION parameter. If *W* is nonzero, the light is positional, and the *(x, y, z)* values specify the location of the light in homogeneous object coordinates. This value is transformed by the model-view matrix, so be careful where and how you define it!

```
// position of light
GLfloat light_position[] = { -1.0, 1.0, 0.0, 1.0 };
glLightfv(GL_LIGHT0, GL_POSITION, light_position);
 gluLookAt(0.,0.,5, 0,0,-100,0.,1.,0.);
```

The color of the light is defined in a manner similar to what we discussed earlier. The entire code can be found in *Example6_4/Example6_4.cpp*. Try changing the position of the light to see the results.

Spotlight

Theater lights or bulbs enclosed in a lamp shade are good examples of spotlights. A spot light is like a point light, but its light rays are restricted to a well defined cone. This kind of light is often used to direct the viewer's eyes to certain parts of the scene. In OpenGL, you can use the GL_SPOT_CUTOFF parameter to specify a light as a spotlight with the specified angle between the axis of the cone and a ray along the edge of the cone. The angle of the cone at the apex is twice this value, as shown in Fig.6.19. No light is emitted beyond the edges of the

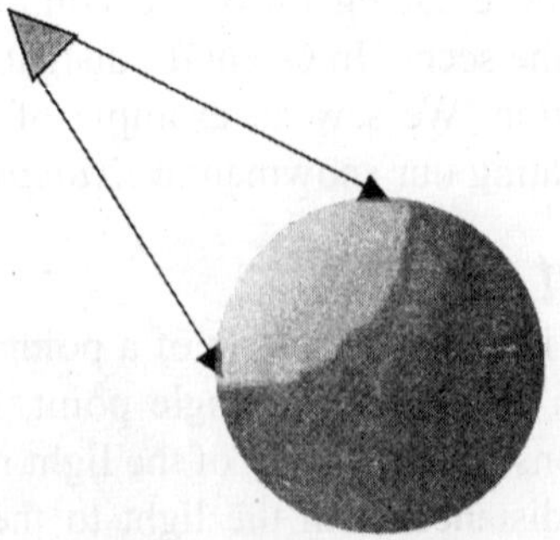

Fig.6.19: A spotlight

cone. The value for GL_SPOT_CUTOFF is restricted to being within the range [0.0,90.0] (unless it has the special value 180.0). The following line sets GL_LIGHT0 to be a spotlight with the cutoff parameter at 20 degrees:

```
glLightf(GL_LIGHT0, GL_SPOT_CUTOFF, 20.0);
```

You also need to specify a spotlight's position and direction, which determines the axis of the cone of light:

```
GLfloat light_position[] = { -1.0, 1.0, 0.0, 1.0 };
GLfloat spot_direction[] = { 0.0, 0.0, -1.0 };
glLightfv(GL_LIGHT0, GL_SPOT_DIRECTION, spot_direction);
glLightfv(GL_LIGHT0, GL_POSITION, light_position);
```

Keep in mind that a spotlight's direction is transformed by the modelview matrix just as though it were a normal vector. An example of a sphere lit by a spot light can be found in *Example6_5/Example6_5.cpp*.

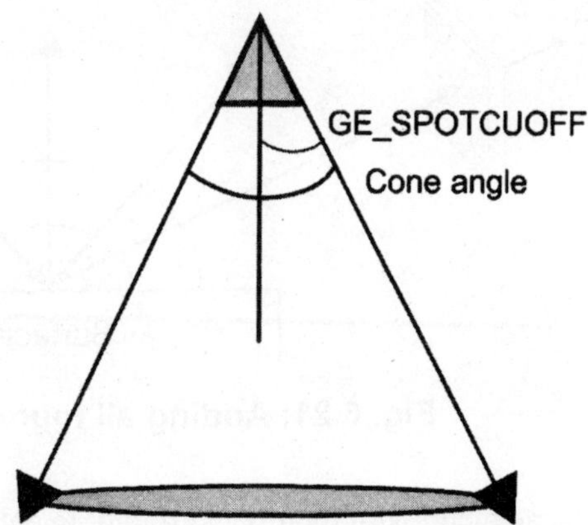

Fig.6.20: GL_SPOT_CUTOFF

Attenuation

For real-world lights, the intensity of light decreases as distance from the light increases, a phenomenon known as attenuation. Since a directional light is infinitely far away, it doesn't make sense to attenuate its intensity over distance, so attenuation is disabled for a directional light. However, you might want to attenuate the light from a positional (point or spot) light. To do this, you can define an attenuation factor to your light. (OpenGL has three different factors for attenuation. Here, we just look at the constant attenuation factor).

```
glLightf(GL_LIGHT0, GL_CONSTANT_ATTENUATION, 1.0);
```

Refer to SHRE03 for more details on attenuation and how to set attenuation factors in OpenGL.

Putting It All Together

The Phong model of CG lighting adds the above four reflectance components to arrive at the color of the light that reflects from the surface. The point of intersection between the reflected light and the image plane determines the pixel being colored.

If there is more than one light source, then the contributions from each light source are added together. The color of the pixel on the screen will be equal to the final color of light that enters the eye/CG camera. The purpose of adding lights to a scene is more than just to enable vision. It is to make the characters in the scene blend with their background, to enhance the mood of the scene, and to make the scene eye catching. In *Example6_6*, we put Snowy against the backdrop of the Alps. (Refer to Color Plate 2.) To make Snowy blend into the scene, we need to identify and match the lighting of the background scene. The scene seems to have a main light coming from the top right hand side. Although

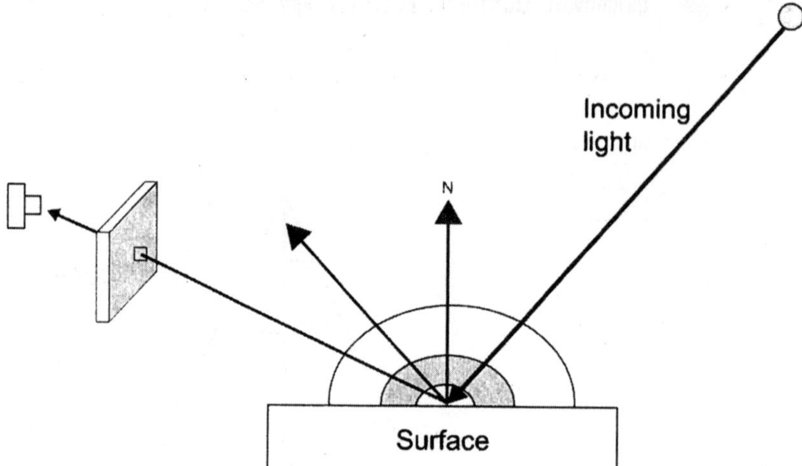

Fig. 6.21: Adding all four components

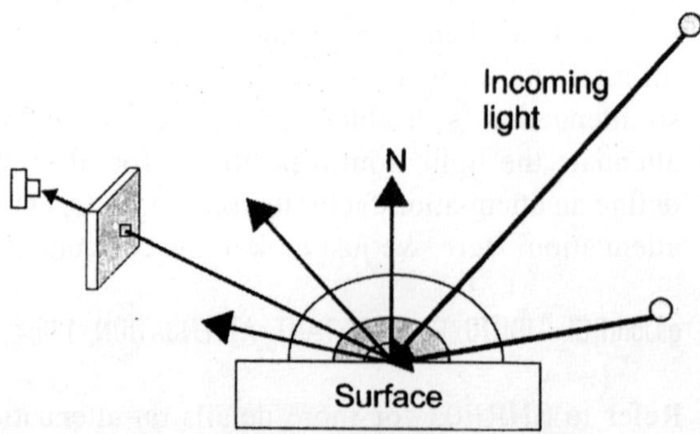

Fig. 6.22: More than one light

this should be yellowish sunlight, the snow is causing it to appear white. A light coming in from the top seems tp enhance the whiteness of the scene. A faint bluish light will probably be reflecting off the snow to light up objects from below.

To light up Snowy and a fake ground that he rests on, we set up three lights for this scene:

```
GLfloat light0_ambient[] = { 0.5, 0.5, 0.5, 1.0 };
GLfloat light0_diffuse[] = { 0.8, 0.8, 0.8, 1.0 };
GLfloat light0_specular[] = { 0.2, 0.2, 0.2, 1.0 };

GLfloat light1_ambient[] = { 0.5, 0.5, 0.5, 1.0 };
GLfloat light1_diffuse[] = { 0.3, 0.3, 0.5, 1.0 };

GLfloat light2_diffuse[] = { 1,1,1, 1.0 };
```

The final color of Snowy as seen in the color plate is the sum of the contributions from ligh0, light1, and light2. The code can be found under *Example6_6*.

6.5 The Normal Vectors

You may have noticed that both diffuse and specular reflection depend on the normal vector to the surface. How and where is this normal vector being defined?

We saw in Chapter 5 that the normal vector to a polygon (or to a vertex on the polygon) determines the orientation of the polygon. This normal vector can be used to determine front- vs. back-facing polygons: polygons: with normal vectors facing away from the viewpoint are back-facing and are culled from the scene, as shown in Fig.6.23. The normal vector also determines the orientation of the polygon or to a vertex on the polygon, relative to the light sources. This normal vector is used in the lighting calculations we looked into earlier. In OpenGL, we can define the normal vector by using the commands

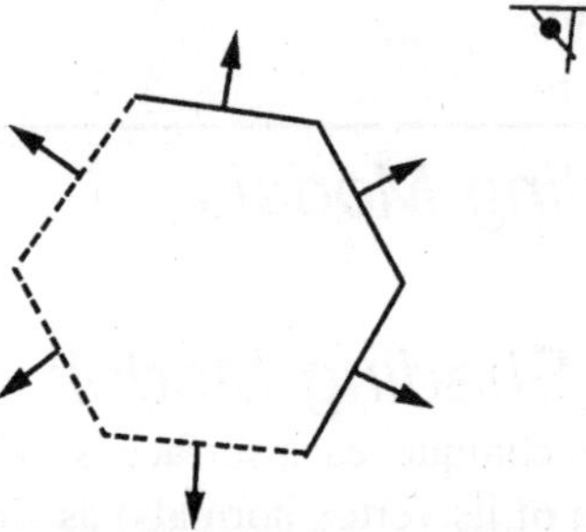

Fig.6.23: Front- and back-facing polygons determined by the direction of normal vectors

```
glNormal3( Glfloat nx, Glfloat ny, Glfloat nz);
void glNormal3fv(const GLfloat *v);
```

This command sets the current normal to the value of the argument passed in. Subsequent calls to **glVertex*()** cause the specified vertices to be assigned the current normal. For the glutSolidObject functions that we use, the normal vectors are already identified within the routine.

In *Example6_7*, we read in the VRML model of the android by using our ReadVRML function. Recall that this function also reads in the normal vectors to the vertices of the polygons being defined.

In the Display function we assign normals to the vertices as shown:

```
glBegin(GL_TRIANGLES);
  glNormal3fv(&(normals[j][3*nindices[j][i]]));
  glVertex3fv(&(coords[j][3*indices[j][i]]));
  glNormal3fv(&(normals[j][3*nindices[j][i + 1]]));
  glVertex3fv(&(coords[j][3*indices[j][i + 1]]));
  glNormal3fv(&(normals[j][3*nindices[j][i + 2]]));
  glVertex3fv(&(coords[j][3*indices[j][i + 2]]));
glEnd();
```

We also assign a shiny yellow material to the android. For each vertex, OpenGL uses the assigned normal to determine how much light that particular vertex receives and reflects, in order to generate the image. Run this program to see the Android in all its metallic glory!

In the last section, we learned how to determine the color of a surface. For every surface point, the four shading components are added together for each light in the scene. This value is then clamped to the maximum intensity of 1 to arrive at the final color of the point.

Performing this calculation for each surface point can be very tedious. Consequently, shading models are used that limit the calculations to certain key points on the surface. The surface is then colored by some average of these values. For polygonal surfaces, the key points are the vertices of the defining polygons.

6.6 Shading Models

Flat Shading Model:

In this technique, each surface is assumed to have one normal vector (usually the average of its vertex normals) as shown in Fig.6.24. This normal vector is used in the lighting calculations, and the resultant color is assigned to the entire surface. In Example6_7, we change the shading model of the Android to be by calling

```
glShadeModel (GL_FLAT);
```

Notice how flat shading causes a sudden transition in color from polygon to polygon. This is because every polygon is rendered using a normal vector that changes abruptly across neighboring polygons. Flat shading is fast to render, and is used to preview rendering effects or on objects that are very small.

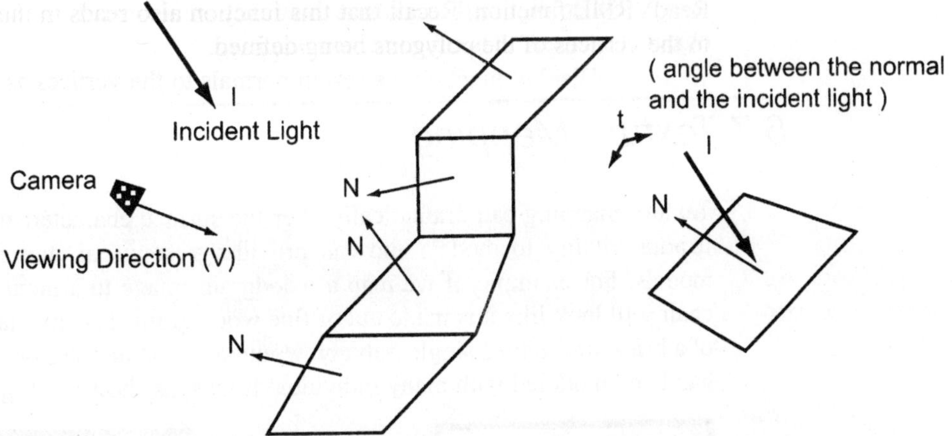

Fig. 6.24: Flat shading: one normal per polygon

Gourad Shading

Gourad shading is used when smooth shading effects are needed. In this process, each vertex defines a normal vector. This normal value is used to calculate the color at every vertex of the surface.

The resulting colors are averaged into the interior of the surface to achieve a smooth render. Since polygons share vertices (and hence the same normal value),

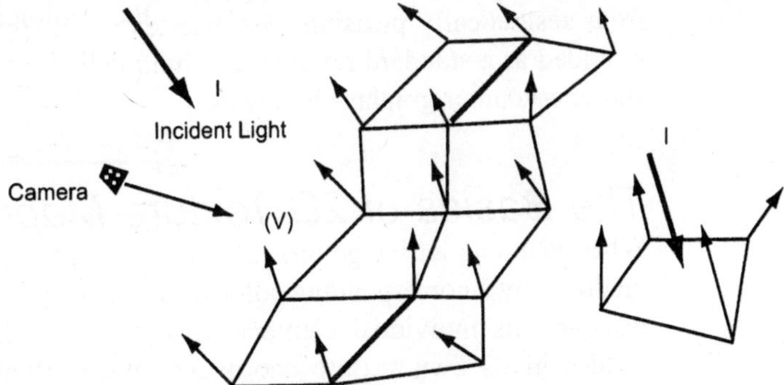

Fig. 6.24: Gourad shading: one normal per vertex

the polygon edges blend together to produce a smooth look. Gourad shading produces smooth renders but takes longer to complete than flat shading. Even more complicated rendering techniques exist, such as ray tracing, radiosity and volume rendering. We shall look into some of these techniques in Chapter 9, when we explore advanced concepts. All techniques a trade-off between computation time and intricacy of the final render. In the last few sections, we saw how to assign material colors to models and then render them with a desired shading scheme. As detail becomes finer and more intricate, explicit modeling becomes less practical. An alternative method often used is to superimpose a completely independent image (either digitized or programmatically generated) over the surface of the object. This method is called *texture mapping.*

6.7 Texture Mapping

Texture mapping can dramatically alter the surface characteristics of an object. It adds vitality to models and can provide great visual cues for even simple models. For example, if we map a woodgrain image to a model of a chair, the chair will look like it is made out of fine wood grain. Texture mapping an image of a brick wall onto a single polygon would give the impression that the polygon had been modeled with many individual bricks, as shown in Fig.6.26.

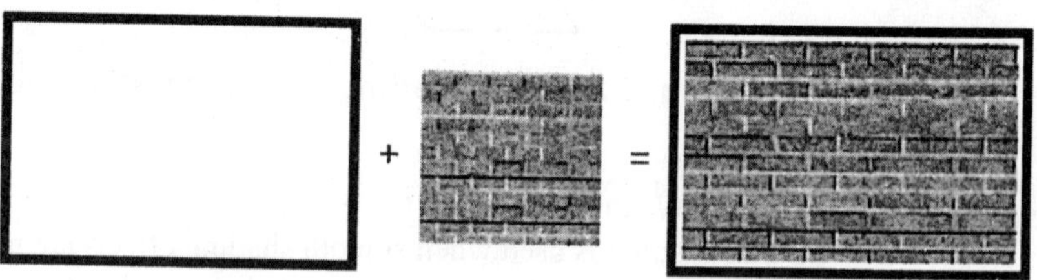

Fig.6.26: Texture mapping a polygon

Texture mapping is a crucial element in today's games and graphic-oriented programs. Without texture mapping, the models that are rendered would be far from aesthetically pleasing. Because texture mapping is so useful, it is being provided as a standard rendering technique both in graphics software interfaces and in computer graphics hardware.

The Basics of 2D Texture Mapping

When mapping an image onto an object, the color of the object at each pixel is modified by a corresponding color from the image. The image is called a *texture map* and its individual elements (pixels) are called *texels*. The texture map resides in its own texture coordinate space, often referred to as (s,t) space. Typically (s,t) range from 0 to 1, defining the lower and upper bounds of the image rectangle.

The simplest version of texture mapping can be accomplished as described. The corner points of the surface element (in world coordinates) is mapped onto the texture coordinate space using a predetermined mapping scheme. The four points in the (s,t) space defines a quadrilateral. The color value for the surface element is calculated by the weighted average of the texels that lie within the quadrilateral. For polygonal surfaces, texture coordinates are calculated at the vertices of the defining polygons. The texture coordinate values within the polygon are linearly averaged across the vertices.

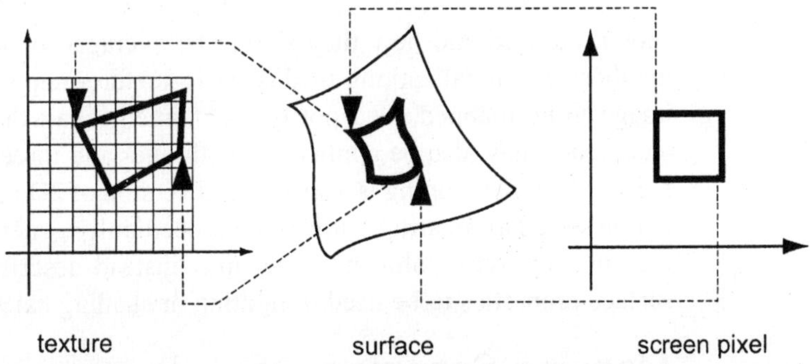

Fig.6.27: Texture mapping: from texels to pixels

The texture space is defined to be from 0 to 1. For texture coordinates outside the range [0,1] you can have the texture data either clamp or repeat over (s,t) as shown in Fig.6.28.

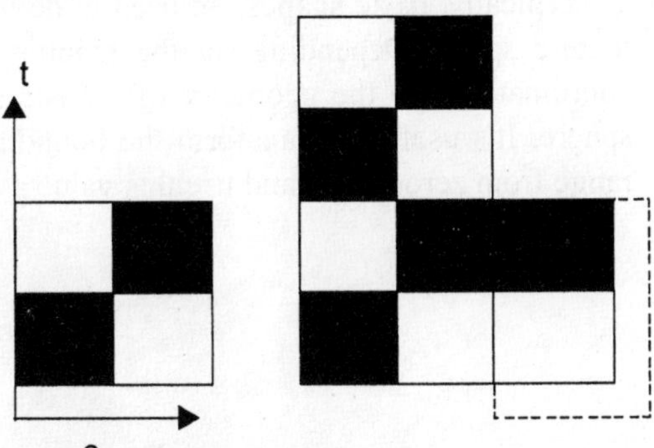

Fig.6.28: A texture image, repeated along t and clamped along s

After being mapped to a polygon or surface and transformed into screen coordinates, the individual texels of a texture rarely correspond to individual pixels of the final screen image. Depending on the transformations used and the texture mapping applied, a single pixel on the screen can correspond to anything from a tiny portion of a texel (magnification) to a large collection of texels (minification), as shown in Fig.29. Filtering operations are used to determine which texel values

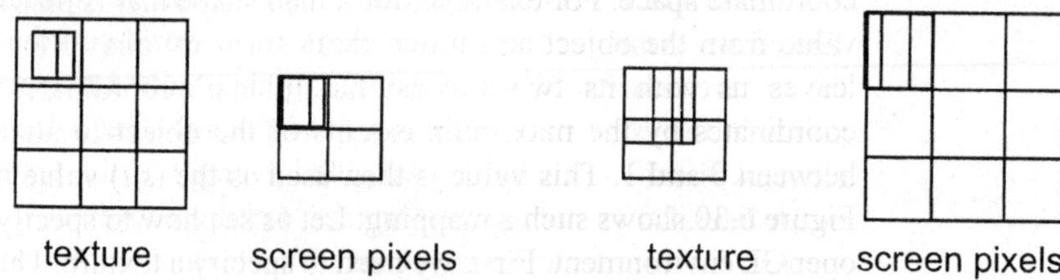

Fig.6.29: Magnification and Minification of texels

should be used and how they should be averaged or interpolated. There are a number of generalizations to this basic texture-mapping scheme. The texture image to be mapped need not be two-dimensional: the sampling and filtering techniques may also be applied for both one- and three-dimensional images. In fact 1D texture mapping is a specialized version of 2D mapping. The texture may not be stored as an array but may be procedurally generated. Finally, the texture may not represent color at all but may instead describe transparency or other surface properties to be used in lighting or shading calculations.

Mapping Schemes

The question in (2D) texture mapping is how to map the two dimensional texture image onto an object. In other words, for each polygonal vertex (or some other surface facet) in an object, we encounter the question, "Where do I have to look in the texture map to find its color?"

Basic Mapping Scheme

Typically, basic shapes are used to define the mapping from world space to texture space. Depending on the mapping situation, we project the object's coordinates onto the geometry of a basic shape such as a plane, a cube, or a sphere. It's useful to transform the bounding geometry so that it's coordinates range from zero to one and use this value as the (s,t) coordinates into the texture

Fig.6.30: Planar Texture mapping: mapping an image to a vase.

Color Plate 1:RGB colors and the resultant colors when these colors are mixed

Color Plate 2: Snowy on the Alps

Color Plate 3: Texture Images used to define a cube mapping

Color Plate 4: Cube mapped objects seem like they are reflecting their environment

Color Plate 5: A Ray traced scene showing reflections and refractions

Color Plate 6: A scene from the game on Mars.

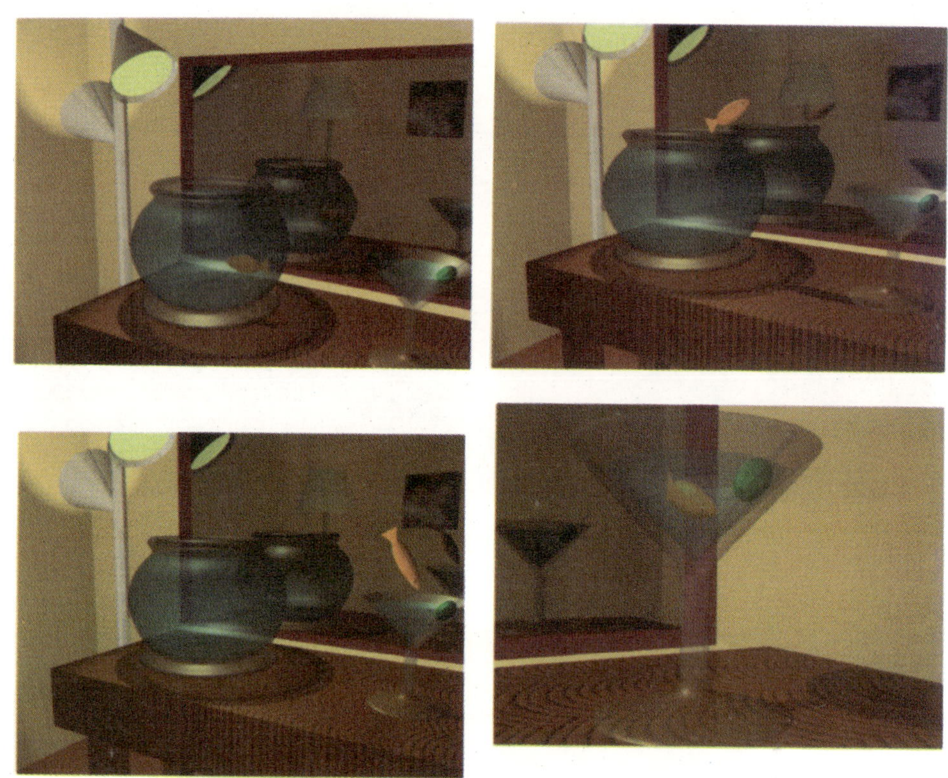

Color Plate 7: Scenes from *A Fish Story*

coordinate space. For example, for a map shape that is planar, we take the *(x,y,z)* value from the object and throw away (project) one of the components, which leaves us with its two-dimensional (planar) coordinates. We normalize the coordinates by the maximum extents of the object to attain coordinate values between 0 and 1. This value is then used as the *(s,t)* value into the texture map. Figure 6.30 shows such a mapping. Let us see how to specify texture maps in our openGL environment. First, we need to specify a texture. The OpenGL command to specify a two-dimensional image as a texture is

```
glTexImage2D()
```

The width and the height of the image *must* be a power of 2. We shall look into the arguments of this command shortly. To set the magnification and minification filters in openGL, you use the command

```
glTexParamateri(GL_TEXTURE_2D, GL_TEXTURE_{MAG/MIN}_FILTER, filter);
```

where the value for *filter* is GL_LINEAR by default, which would define a linearly interpolated filter. GL_NEAREST takes the closest pixel in the texture image rather than using interpolation, and hence is very quick to render. As noted earlier, texture coordinates lie in the 0-1 range. When the coordinates go out of this range, we would like the texture to either be clamped at the endpoints or repeated. You can specify the wrapping of the texture in both *s* and *t* independently by

```
glTexParamteri(GL_TEXTURE_2D, GL_TEXTURE_WRAP_{S/T}, wrap)
```

where *wrap* can be GL_CLAMP or GL_REPEAT.

Finally you specify the texture coordinates at the current vertex location by making the call

```
glTexCoord2d(s,t)
```

right before you make the glVertex call. Note that all 2D texture commands have a corresponding 1D call. You also need to define the texture mode to be used while rendering. GL_MODULATE modulates the current lighting and color information with the texture image. GL_DECAL uses only the texture color. You can set the texture mode using the glTexEnvi function:

```
glTexEnvi(GL_TEXTURE_ENV, GL_TEXTURE_MODE, mode);
```

Let us work through an example of applying a planar texture on three kinds of objects: a vase, a sphere, and a cube. The vase is a VRML model that we downloaded from the Internet.

Assuming projection along the *z*-axis, the mathematics to calculate the planar texture coordinates is simple. Given any vertex point (*x,y,z*) and the object's *x* and *y* extents (XMIN, XMAX) and (YMIN, YMAX); we can define a function to map the world coordinates to texture coordinates as:

```
void MapCoordinates(GLfloat x,GLfloat y, GLfloat z, int mapping)
{
	glTexCoord2f((x-XMIN)/(XMAX-XMIN),(y-YMIN)/(YMAX- YMIN));
}
```

Now, before defining the polygon vertices, we make a call to this function to set the corresponding texture coordinates. For example, we define the vertices of our VRML object as follows:

```
MapCoordinates(coords[3*indices[i]], coords[3*indices[i] + 1], coords[3*indices[i] + 2], mapping);
glNormal3fv(&(normals[3*nindices[i]]));
glVertex3fv(&(coords[3*indices[i]]));
```

In *Example6_8*, we define the planar texture for the three models. In the init() function, we first generate and bind a new texture:

```
glGenTextures(1, &texName);
glBindTexture(GL_TEXTURE_2D, texName);
```

We then set the various wrap and filter parameters for this texture:

```
glTexParameteri(GL_TEXTURE_2D,GL_TEXTURE_WRAP_S, GL_CLAMP);
glTexParameteri(GL_TEXTURE_2D,GL_TEXTURE_WRAP_T, GL_CLAMP);
glTexParameteri(GL_TEXTURE_2D, GL_TEXTURE_MAG_FILTER, GL_NEAREST);
glTexParameteri(GL_TEXTURE_2D, GL_TEXTURE_MIN_FILTER, GL_NEAREST)
```

Next, we actually read in the texture image—which is nothing more than a BMP image file. The image size is a power of 2, namely, 512 by 512

```
bits - ReadBitmap("checktexture.bmp", &info);
```

We define the loaded bitmap to be the current texture map.

```
glTexImage2D(GL_TEXTURE_2D, 0, 3, info->bmiHeader.biWidth,info->bmiHeader.biHeight,0,
GL_BGR_EXT, GL_UNSIGNED_BYTE,bits);
```

Finally we enable texture mapping and the texture environment.

```
glEnable(GL_TEXTURE_2D);
glTexEnvf(GL_TEXTURE_ENV,GL_TEXTURE_ENV_MODE, GL_MODULATE);
```

The *Display* routine calls the drawing routines for the texture mapped objects. The complete example can be found under *Example6_8* in files *Example6_8.cpp, Shapetextures.cpp* and *Shapes.h*. You will need to compile the program with the vrml.cpp, vrml.h, bmp.cpp and bmp.h files provided as well. The texture image can be found under the *Images* directory.

Other bounding shapes can be used to make the mapping work more like a shrink-wrap. Refer to [FOLE95] for more information on other mapping schemes.

Environment Mapping

If you look around your room you may realize that many objects reflect their surroundings. A bottle of water, a mobile phone, a CD cover, a picture frame, etc. are only a few examples of reflecting objects that could be found in any 3D scene. To make the 3D world more realistic, objects should show reflections of their environment. This reflection is often achieved by a texture-mapping method called *environment mapping*.

The goal of environment mapping is to render an object as if it were reflective, so that the colors on its surface are those reflected from its surroundings. In other words, if you were to look at a perfectly polished, perfectly reflective silver object in a room, you would see the walls, floor, and other objects in the room reflected off the object. (A classic example of using environment mapping is the evil, morphing *Cyborg* in the film *Terminator 2*.) The objects whose reflections you see depend on the position of your eye and on the position and surface angles of the silver object. Of course, objects are not usually completely reflective, so the color of the reflection is modulated with the object's actual color for a realistic look.

True environment mapping can be achieved using ray tracing, a technique we shall learn about in Chapter 7. Ray tracing is a very expensive option and is usually not necessary for most cases. More often, certain tricks are used to

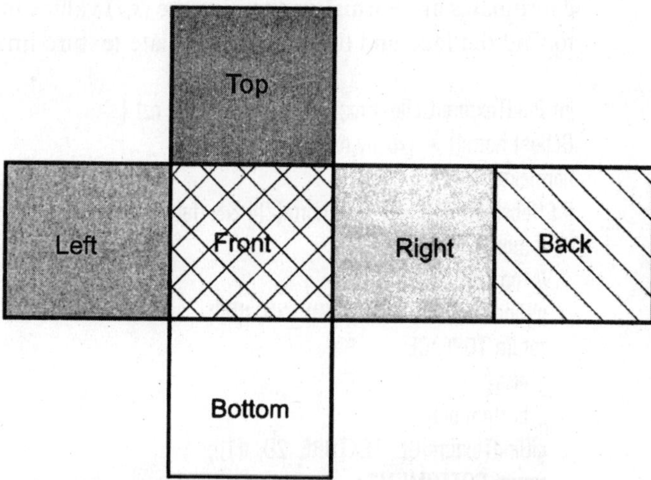

Fig.6.31: Six textures for cube mapping

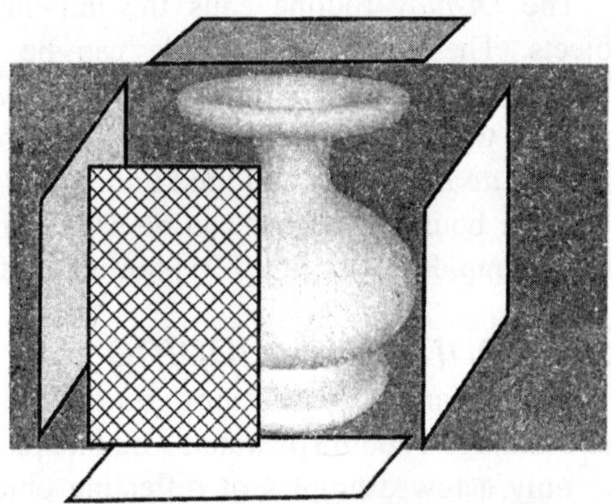

Fig.6.32: Cube Mapping

achieve reflective effects. A method often employed is called cube environment mapping. The idea behind cube environment mapping is very simple. From an object with a reflective surface you first generate six images of the environment in each direction (front, back, up, down, left, right). That is, imagine placing a camera at the center of the object and taking the size photographs of the world around it in the directions specified. Based on the normal vector of the vertex in consideration, the appropriate image is selected. This image is then projected using planar mapping onto the object as shown in Fig.6.32.

The resultant object looks like it is reflecting its environment!

Mathematically, the cube face is identified by using the normal vectors (*nx,ny,nz*) on the surface of the object. The greatest component is used to identify where the surface should be "looking"; and hence the cube face and the texture image to be used. The other two coordinates are used to select the texel from the texture by a simple 3D to 2D projection. If, say, *ny* was the highest value component then we divide the other components by *ny* (*nx/ny,nz/ny*). These coordinates are normalized to give the (*s,t*) values into the texture map. The code to find the face and bind the appropriate texture image is shown below:

```
int BindTexture(GLfloat nx, GLfloat ny, GLfloat nz) {
GLfloat norm[] = {nx,ny,nz};
normalize(norm);
if ( (fabs(norm[1]) >= fabs(norm[2])) && (fabs(norm[1])>= fabs(norm[0]))){
   if (norm[1] >=0.) {
   // top only
   glBindTexture(GL_TEXTURE_2D, t[0]);
   return TOPFACE;
   } else{
   // bottom only
   glBindTexture(GL_TEXTURE_2D, t[1]);
   return BOTTOMFACE;
```

```
  }
 } else if (fabs(norm[2])> fabs(norm[0])){
   if (norm[2] > -0.) {
     glBindTexture(GL_TEXTURE_2D, t[2]);
     return FRONTFACE;
   } else {
     glBindTexture(GL_TEXTURE_2D, t[3]);
     return BACKFACE;
   }
 } else
 {
   if (norm[0]> -0.) {
     glBindTexture(GL_TEXTURE_2D, t[4]);
     return RTFACE;
   }
   else{
     glBindTexture(GL_TEXTURE_2D, t[5]);
     return LTFACE;
   }
 }
 return -1;
}
```

The code to use planar mapping to texture-map the coordinates is as follows:

```
void MapCoordinates(GLfloat nx,GLfloat ny, GLfloat nz, int face)
{
 if (face - - TOPFACE)
   glTexCoord2f((nx/ny + 1)/2.,(1-nz/ny)/2.);
 else if (face - - BOTTOMFACE)
   glTexCoord2f((1-(nx/ny))/2.,(1-(nz/ny))/2.);
 else if (face - - FRONTFACE)
   glTexCoord2f((nx/nz + 1)/2.,(ny/nz + 1)/2.);
 else if (face - - BACKFACE)
   glTexCoord2f((1 + nx/nz)/2.,(1-ny/nz)/2.);
 else if (face - - RTFACE)
   glTexCoord2f((1-(nz/nx))/2.,(ny/nx + 1)/2.);
 else
   glTexCoord2f((-(nz/nx) + 1)/2.,(-(ny/nx) + 1)/2.);
}
```

Before making a call to begin a polygon, we first find the face of the cube that this polygon will index and then call MapCoordinates to generate the correct (s,t) value into the texture. In *Example6_9*, we read in the six texture images shown in Color Plate 3 as the six faces of the cube, and use them to environment map the vase, sphere and cube. The final code can be found under *Example6_9*. You will also need *vrml.cpp, vrml.h, bmp.cpp, bmp.h* and *utils.h* to compile and run this program. The textures can be found under the *Images* folder under the installed directory for the sample code. The rendered objects are shown in Color Plate 4.

All the examples above use the vertex/normal data in object space to map into the texture. This means that if our object moves in the 3D space, the texture will be 'stuck' onto it. This may not be desirable in the case of environment mapping. As the object moves in the world, you want the reflections to change accordingly. In order for the map to move, you need to pass in the vertex/normal data modified by the current transformation space. Use the OpenGL command

```
glGetFloatv(GL_MODELVIEW_MATRIX , modelview);
```

to get the current model view matrix and apply the necessary matrix operations on your vertex/normal data. Interested readers should try using this setup to map the cube environment onto our marching android from Chapter 5.

6.8 Vertex Shaders

No discussion on rendering would be complete without talking about *vertex shaders*. The concept of shaders is nothing new to the graphics industry, Pixar has used it for years in their phenomenal films such as *Toy Story* and *The Incredibles*. However till now, vertex shading effects were so computationally complex that they could only be processed offline using server farms. With the current generation of semiconductor technology, developers can program real-time Vertex Shaders to breathe life and personality into characters and environments, such as fog that dips into a valley and curls over a hill; or true-to-life facial animation such as dimples or wrinkles that appear when a character smiles.

A vertex shader is a graphics processing function used to add special effects to objects in a 3D environment by performing mathematical operations on the object's vertex data. Vertex data refers to data of the points along the surface, at some specified precision (and not just at the vertices of the defining polygons!) Each vertex data is defined by many variables. For instance, a vertex is always defined by its location in a 3D environment using the *x*-, *y*-, and *z*- coordinates. Vertices may also be defined by their normal vector, color, texture, and lighting characteristics. Vertex Shaders don't actually change the type of data; they simply change the values of the data, so that a vertex emerges with a different color, different textures, or a different position in space, and is rendered as such on the screen.

A vertex shader can be thought to be a magic box. Vertex data goes in and

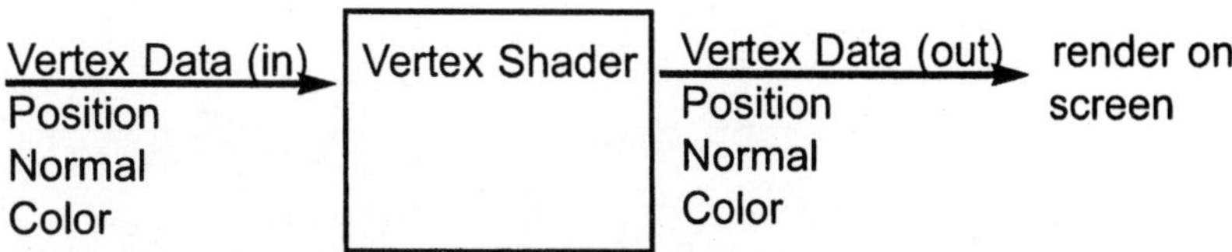

Fig.6.33: Vertex shader

modified vertex data comes out. What the box does depends on how the programmer develops the shader.

With the release of OpenGL1.4, the OpenGL group has released the *OpenGL Shading Language*. The OpenGL Shading Language has been designed to enable programmers to express vertex shaders in a high level language.

Independently compiled programs that are written in this language are called *shaders*. A program is a set of shaders that are compiled and linked together. OpenGL provides entry points to manipulate and communicate with these shaders and apply them to objects within the CG world being defined for stunning looking renders.

Vertex shaders is a vast topic, and deserves an entire book to do it full justice. Refer to [ROST04] for more information on this topic. We are confident that with the concept you have learned in this book, vertex shaders will not be difficult for you to grasp.

Summary

In this chapter, we have studied the important techniques used in rendering. Our quest for visual realism has taken us from hidden surface removal to lighting and materials to texture mapping of surfaces. These techniques approximate the physics of sight to generate photorealistic images. We shall look into some more advanced rendering algorithms in the next chapter.

Chapter 7

Advanced Techniques

Graphics scenes can contain different kinds of objects: trees, flowers, glass, fire, water, etc. So it is not too surprising that numerous techniques can be applied to depict the specific characteristics of each object. In the last few chapters, we learned about the basic principles of modeling and rendering. In this chapter, we shall look into some advanced techniques popular in the CG world.

In Chapter 5, we dealt with linear components (lines and polygons) to depict the hull of models. For complex models, polygonal surfaces can be tedious to build for and may also occupy large amounts of storage space. Curves and surfaces are often used instead in order to achieve smoother representations with less storage space. In this chapter, we look into several kinds of curves and surfaces used in CG. Subdivision surfaces is another technique that has become popular to achieve a high degree of local refinement.

In Chapter 6, we saw some basic rendering techniques. Although we went from wire-frame to real shaded images, we are still not close to the photo-realistic images that we set out to produce. Our images don't have shadows or transparency, for example. In this chapter, we shall also talk about an advanced algorithm called *ray tracing* to achieve such effects. *Radiosity* is another rendering technique that we shall discuss briefly.

In this chapter, you will learn

- Advanced modeling
  - Curves—Bezier, B-splines, and Nurbs
  - Nurbs surfaces
  - Other techniques
- Advanced rendering algorithms
  - Ray tracing
  - Radiosity

7.1 Advanced Modeling

Numerous techniques are used to model objects in CG. Which technique is used depends on the object being modeled, as well as the final outcome desired. Polygonal models are quick and easy to display but do not necessarily depict a smooth surface. They are used widely in the gaming industry for speed of display. Other techniques can depict smooth hulls but are slower to display. One such technique is the use of parametric curves and surfaces.

Parametric Curves

Curves are used in CG for all kinds of purposes—fonts, animation paths and modeling, to name a few. It is instructive to look into the mathematical equations used to represent curves to understand how they are used in CG.

Parametric Equations

If you recall high school algebra, the implicit function for a curve in 2D space is of the type:

$$f(x,y) = 0$$

For example, a circle with radius r centered at 0 has a defining equation

$$x^2 + y^2 - r^2 = 0$$

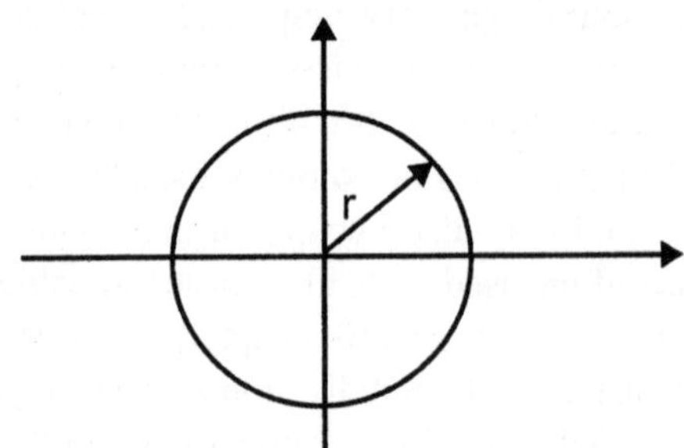

Fig.7.1: Curve segments meeting at a join point

All points (x,y) that satisfy the above equation are along the path of the circle.

A line can be thought of as a specialized curve—it's a linear curve! A line with two endpoints, one at $(x1,y1)$ and the other at $(x2,y2)$ has the implicit equation

$$y - m*(x - x1) - y1 = 0$$

where $m = (y2-y1)/(x2-x1)$ and is also referred to as the slope of the line.

All points, (x,y) that satisfy the above equation are positioned on the line being defined. The same curve/line can also be represented using a parametric function. Parametric functions define each coordinate against a parameter of choice. We saw in Chapter 1, that our circle can be represented against the parameter θ as

$$x = r\ cos\theta,\ \ y = r\ sin\theta$$

As θ vary's from *0* to 360, we *evaluate* the points along the circle.

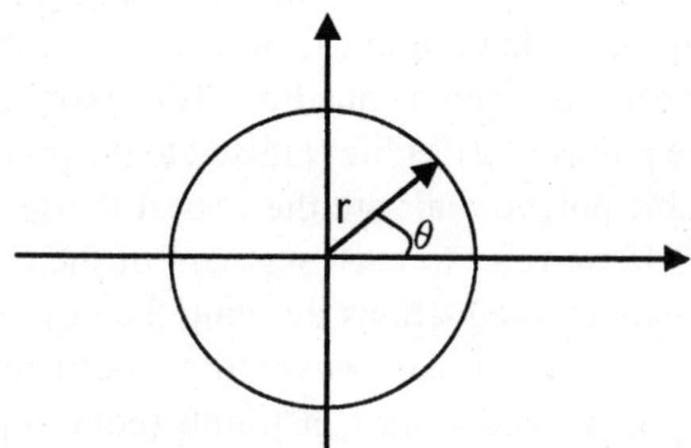

Fig.7.2: Varying θ from 0 to 360

Our line can also be defined in a parametric form against a parameter t as

$x = (x2 - x1)*t + x1$

$y = (y2 - y1)*t + y1$

as t varies from 0 (defining the endpoint $(x1,y1)$ to 1 (defining the endpoint$(x2,y2)$) we *evaluate* the points along the line. It can be shown that in a 3D space, we can define a line against a parameter t as

$x = a_x t + d_x$

$y = a_y t + d_y$

$z = a_z t + d_z$

At t =0, we attain the endpoint of the line (d_x, d_y, d_z) and at t=1 we attain the endpoint $(a_x+d_x,\ a_y+d_y,\ a_z+d_z)$. These two end points are enough to define the entire line. In other words, they define the constraints of the line, and are also called the control points of the line.

In Chapter 1, we saw how to use a set of lines (line segments) to approximate the representation of a curved object—the circle. In CG, we say that the curve (the circle in this case) is being approximated by linear segments. The segments are called *linear*, because they are defined by an equation which is linear in parameter t (t raised to the power of 1). In Chapter 5, we saw how a set of polygons can approximate a curved surface such as a sphere—a linear approximation to a curved surface.

Linear approximations to a curve (or a surface) usually requires a large amount of vertex and normal data for reasonable closeness to the smooth curve/surface. Defining a large number of vertices to approximate smooth curves and surfaces is not only tedious but also error prone and difficult to manipulate.

Pierre Bezier was the first to develop a set of parametric cubic equations to represent curves and surfaces using only a small set of control points. Many more functions have been developed that produce better results at the cost of efficiency. Keep in mind that these functions still only approximate the surface/curve, but do so with less storage and offer easier manipulation than their linear counterparts.

Cubic Curves

Curves can be approximated by a set of segments. We have seen how to use of segments that are linear in nature. But the approximating segments need not be linear. They can be quadratic (the defining parametric polynomials are t raised to the power of 2), cubic (t raised to the power 3), or even higher orders. In practice, cubic polynomials are the easiest to use and control and are very popular in CG.

If we refer to each segment of the curve as $Q(t) = (x(t), y(t), z(t))$, the cubic polynomial equations defining the (x,y,z) points along the curve are of the type: Cubic polynomials have four coefficients (a,b,c and d), and hence a curve segment needs four constraints (control points) to define it.

$$x(t) = a_x t^3 + b_x t^2 + c_x t + d_x$$
$$y(t) = a_y t^3 + b_y t^2 + c_y t + d_y$$
$$z(t) = a_z t^3 + b_z t^2 + c_z t + d_z$$
$$(0 \leq t \leq 1)$$

The different kinds of curves (and surfaces) are classified by the coefficient values of the polynomial equation. The coefficient values determine how the curve *interpolates* the control points. Some of the commonly used curves are Bezier, B-Spline, Hermite, etc. The two kinds of curves we shall study in this book are *Bezier* curves and *B-splines* (specifically Nurbs).

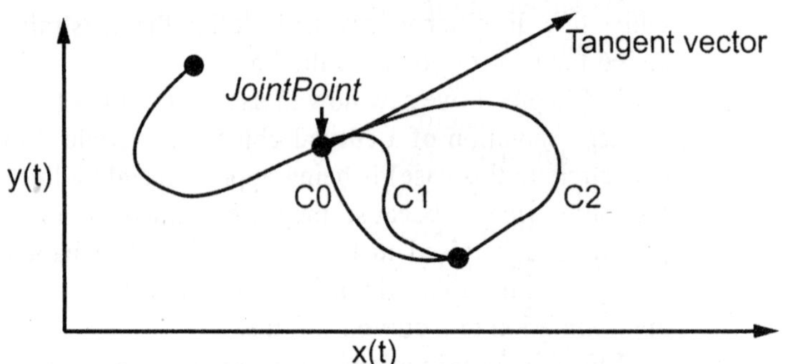

Fig.7.3: Curve segments meeting at a join point

Since a curve is approximated by more than one curve segment, it is important to understand how the curve segments behave at the point at which they meet which is called a *join point*. If two curve segments of a curve meet together, the curve is said to have *C0 continuity*. Usually, a curve should at least be C0 continuous to define a continuous curve! If in addition the tangent vectors of the two curve segments (the first derivative, also called as velocity of the curve) are equal at the join point, the curve is said to be *C1 continuous*. If the curvature of the two curve segments (second derivative, also called rate of change or acceleration of the curve) are also equal at the join point, then the curve is said to be *C2 continuos* as shown in Fig.7.3. C1 continuity is required in curves to guarantee a smooth curve. C2 continuity guarantees a smooth transition between segments.

Splines and, in particular, Nurbs have C1 and C2 continuity at their join points and are often preferred over other types of curves.

Bezier Curves

The (cubic) Bezier curve segment has four control points: two control points, P_1 and P_4, define the endpoints of the curve, and the two control points, P_2 and P_3, affect the shape of the curve by controlling the end tangent vectors.

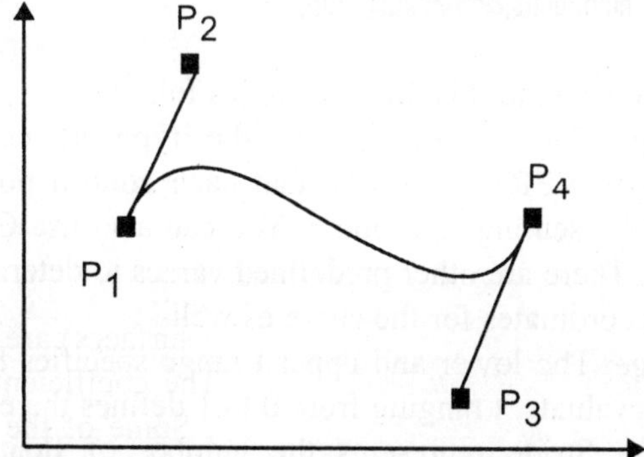

Fig.7.4: Bezier curve defined by four control points

As shown in Fig.7.4, the starting and end tangents are determined by vectors P_1P_2 and P_3P_4. We say that the Bezier curve interpolates (passes through) the first and last control point and approximates the other two. Let us look into how we can define and display Bezier curves using OpenGL.

The OpenGL evaluator functions allow you to use a polynomial equation to produce vertices, normals, texture coordinates, and colors. These calculated values are then passed on to the processing pipeline as if they had been directly specified. The evaluator functions define a Bezier curve and are also the basis for the NURBS (Non-Uniform Rational B-Spline) functions. The first step in using an evaluator is to define the polynomial map by using the OpenGL command

glMap1f

In *Example7_1*,we define 4 control points for the Bezier segment, as

```
GLfloat ctrlPoints[4][3] -
        {-6.,0.,0.,
        -4.,6.,0.,
        4.,-6.,0.,
```

```
6.,0.,0.};
int numPoints - 4;
```

The control points in the cubic map are defined by:

```
glMap1f(
GL_MAP1_VERTEX_3, // target, type of control data, 3 coordinates for each vertex
0.,                 // lower t range
1.,                 // upper t range
3,                  // stride
numPoints,&ctrlPoints[0][0]);
```

The parameters to this function are as follows:

target: The target defines the type of control points being used. GL_MAP1_VERTEX_3 says that each control point has three floating-point values representing *x, y,* and *z*. You can also use GL_MAP1_VERTEX_4 (for x,y,z,w). There are other predefined values to determine the normal vectors and texture coordinates for the curve as well.

trange: The lower and upper t range specifies how much of the curve you wish to evaluate. t ranging from 0 to 1 defines the entire range of the curve.

stride: Stride represents the number of floats or doubles between the beginning of one control point and the beginning of the next one in the data structure referenced in ctrl *points*. This allows control points to be embedded in arbitrary data structures. The only constraint is that the values for a particular control point must occupy contiguous memory locations.

numpoints, ctrlPoints: Finally, we define the number of control points and the array containing the control points.

We need to enable openGL to calculate the polynomial being mapped. We do this by calling

```
glEnable(GL_MAP1_VERTEX_3);
```

The next step is to determine the domain values to be used to evaluate the polynomial. OpenGL works with the concept of grids to specify evenly spaced domain values for evaluation. The command

```
//map a grid of 20 points along the t domain ranging from 0 to 1
glMapGrid1d(20.,0.,1.0);
```

tells openGL to lay a grid of 20 evenly spaced points along the *t* domain. Finally, we need to instruct OpenGL to evaluate the curve at the defined grid points. This is accomplished by the command

```
glEvalMesh1(GL_LINE,0,20);
```

All the evaluated points are then smoothly connected using the primitive specified (in this case, a line). You also need to specify the first and last integer values for grid domain variable t. In this case, the curve will be evaluated and plotted at t =(0,1/20,2/2- … 1).

The more the number of points in our grid, the smoother will be the final curve that is output. When you compile and run *Example 7_1/Example7_1.cpp*, you will see the Bezier curve, defined by the control points specified.

Try changing the number of grid points to see how this affects the smoothness of the curve. Note that this only changes how many points are used to evaluate the polynomial equation—it doesn't change the equation being used.

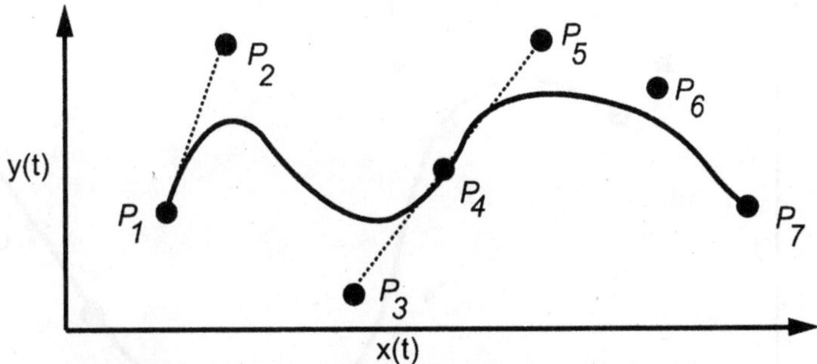

Fig.7.5: Piecing together Bezier curve segments

Two Bezier curve segments can be constructed to have C0 and C1 continuity if we can ensure that they share a join point (P_4) and the tangent vectors at P_4 are equal or $P_5 - P_4 = P_4 - P_3$ as shown in Fig.7.5. Try to define another Bezier segment and connect it to the segment we defined in *Example7_1*. How will you define the control points to achieve C0 continuity? To achieve C1 and C2 continuity?

One can approximate entire curves by assembling Bezier curve segments in this manner. For more complex curves, the mathematics to ensure C1 continuity across the curve while adjusting control points gets tricky. A better solution is provided by splines and by what we call Nurbs.

B-Splines

Spline curves originated from flexible strips used to create smooth curves in traditional drafting applications. Much like Bezier curves, they are formed mathematically from piecewise approximations of cubic polynomial functions.

B-Splines are one type of spline that is perhaps the most popular in computer graphics applications. The control points for the entire B-spline curve are defined in conjunction. They define C2 continuous curves, but the individual curve segments need not pass through the defining control points. Adjacent segments share control points, which is how the continuity conditions are imposed. For this reason, when we discuss splines, we discuss the entire curve (consisting of its curve segments) rather than its individual segments which must then be stitched together. Cubic B-splines are defined by a series of (m=n+1) control points: $P_0, P_1 ... P_n$

Each curve segment of the spline Q_i $3 \leq i \leq n$ is defined by the four control points $P_{i-3}, P_{i-2}, P_{i-1}, P_i$.

For example, in Fig.7.6, we show a spline with m=8 control points. The individual segments of the curve are Q_3, Q_4, Q_5, Q_6 and Q_7. Q_3 is defined by the four control points, P_0-P_3,Q_4 by points P_1-P_4 etc.

Conversely, every control point affects four segments. For example, in the

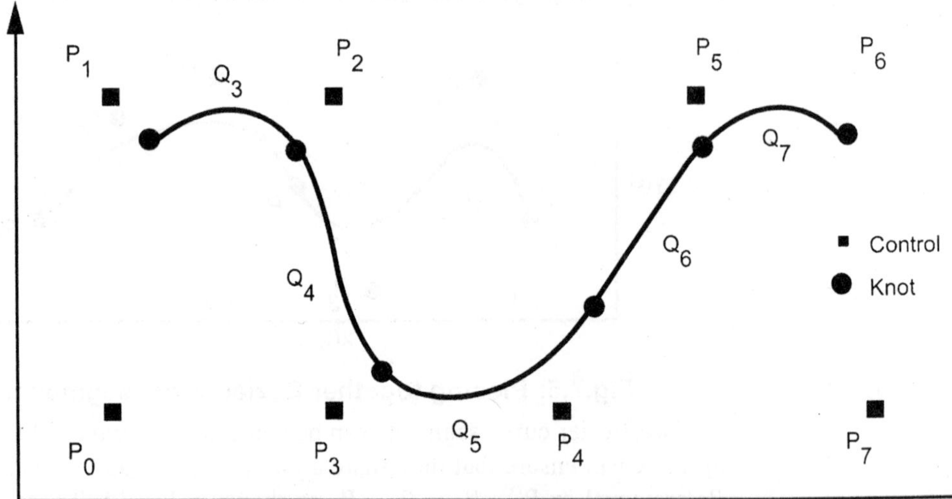

Fig.7.6: A uniform nonrational B-spline

above figure, point P_4 affects segments Q_3, Q_4, Q_5, and Q_6. Moving a control point will affect these four segments but will not affect the entire curve. This is a very useful property of B-splines that we shall look into more in the next few chapters.

The join points between the segments are called *knots*. The knot between segment i and i+1 is represented as k_i. The initial point of the first segment and the endpoint of the last segment are also called knots, so there is a total of (n-1) knots for the spline under consideration. When knots are uniformly placed, as shown in Fig.7.6, the curve spline is called a *uniform non-rational spline*. Unfortunately, it is difficult to define and control the splines since the segments do not interpolate the control points.

Non-uniform non-rational B-splines define (n+5) knots. The knots need not be uniformly spaced—and in fact are user defined. The advantage is that we can force the curve to interpolate certain control points. Non-uniform B-splines uses the notion of knot value sequences: a non-decreasing sequence of knot values that defines the placement of knots of the curve. For example, if we assumed the curve above was a non-uniform non rational B spline, its knot sequence would effectively be (0,1,2,3,4,5,6,7,8,9,10,11). (n+5 = 12 knots)

If successive knot values are equal in the sequence, it is called a *multiple knot*. Multiple knots causes the curve to approximate the associated control point more closely. In fact, three successive knot values forces the curve to actually interpolate the control point, thereby making the shape of the curve easier to define. Defining multiple knots does lead to a loss in continuity, but only at the associated control point. If we modified our curve above to assume a different knot sequence, the results would appear as shown in Fig.7.7.

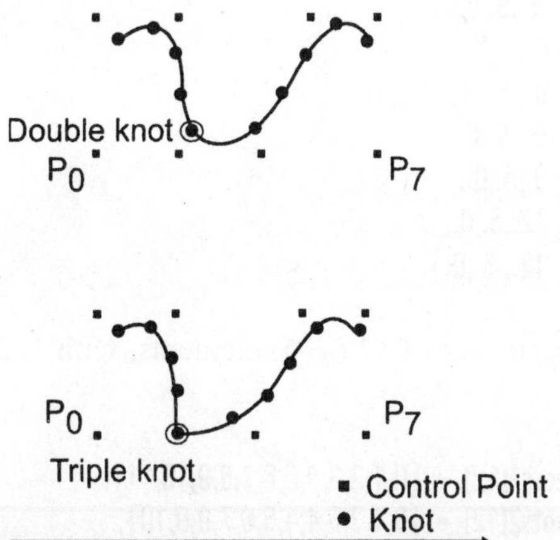

Fig.7.7:Non-uniform non-rational B-splines. Multiple knots, double knot with knot sequence (0,1,2,3,4,4,5,6,7,8,9,10) causes C1 continuity, whereas a triple knot with knot sequence (0,1,2,3,4,4,4,5,6,7,8,9) results in only C0 continuity.

Nurbs

B-splines (or any other non-rational curve) can be defined in homogenous coordinates by adding W(t)=1 as the fourth element in the parametric equation; $Q(t) = (X(t), Y(t), Z(t), W(t))$. (As usual, moving from homogenous coordinates to 3D space involves dividing by $W(t)$). This process is called *rationalizing* the curve. The advantage of rational curves is that they are invariant under rotation, scaling, translation and perspective transformations. This means we need to apply the transformations only to the control points and then re-evaluate the curve equation to generate the transformed curve. A non-uniform rational B-spline is also called a Nurb and is heavily used in CG due to the properties described above.

In *Example7_2*, we draw a Nurbs curve using OpenGL. We first need to create a new Nurbs renderer before we can define the actual curve. This is done by using the command

```
pNurb = gluNewNurbsRenderer();
```

We can set properties of this nurb as follows

```
gluNurbsProperty(pNurb, GLU_SAMPLING_TOLERANCE, 25.f);
gluNurbsProperty(pNurb, GLU_DISPLAY_MODE, GLU_LINE);
```

This property means that the resultant Nurbs curve will be drawn using a line primitive. We define a set of eight control points (n= 7)

```
GLfloat ctrlPoints[8][3] =
        {-5.,-5.,0.,
        -5.,5.,0.,
        0.,5.,0.,
        0.,-5.,0.,
        5.,-5.,0.,
        8.,5.,0.,
        12.,5.,0.,
        12.,-5.,0.};
```

and knot value sequences of 12 (n+5) elements, with 1, 2, and 3 multiple knots, respectively.

```
GLfloat Knots1[12] = {0,1,2,3,4,5,6,7,8,9,10,11};
GLfloat Knots2[12] = {0,1,2,3,4,4,5,6,7,8,9,10};
GLfloat Knots3[12] = {0,1,2,3,4,4,4,5,6,7,8,9};
```

The OpenGL command to define and display a Nurbs curve is

```
gluNurbsCurve
```

The parameters for this function are similar to the parameters we set for glMap but includes support for defining the knot sequence.To define the curve, we bracket it between gluBegnCurve and gluEndCurve.

```
gluBeginCurve(pNurb);
gluNurbsCurve(pNurb,
        12,                     // the number of knots
        Knots1,                 // the knot value sequence
        3,                      //stride
        &ctrlPoints[0][0],      // the control points
        4,                      // the order of the curve
        GL_MAP1_VERTEX_3);      // type of vertex data
gluEndCurve(pNurb);
```

The order of the NURBS curve equals the degree of the parametric equation+1; hence a cubic curve has an order of 4. The entire code can be found in *Example7_2/Example7_2.cpp*. Try experimenting with different knot value sequences to see what results you get. Interested readers are encouraged to refer to **DEB001** for more details on splines.

Parametric Bicubic Surfaces

Parametric bicubic surfaces are a generalization of the cubic curves. The surface is partitioned into parametric surface segments or *patches*, which are stitched together to form the entire surface.

The parametric equation for each patch is defined along two parameter domains – *s* and *t*, defining a surface as opposed to a curve.

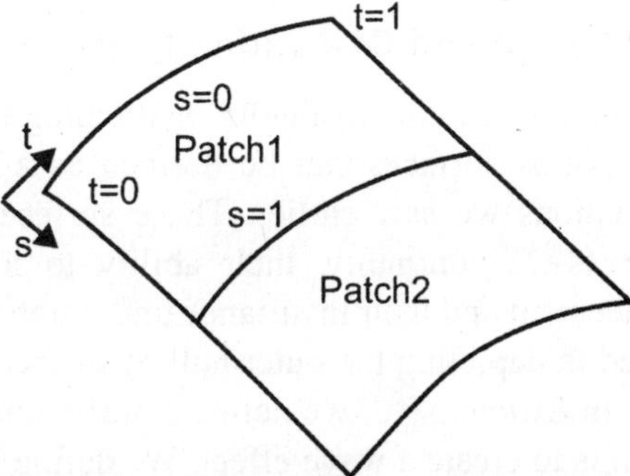

Fig.7.8: C0 and C1 continuity at surface boundary.

Each patch is defined by blending the control points (4 *4 = 16 control points are needed to define a patch). In fact, you can think of the control points defining curves at specific domain intervals that are then meshed together to define a surface.

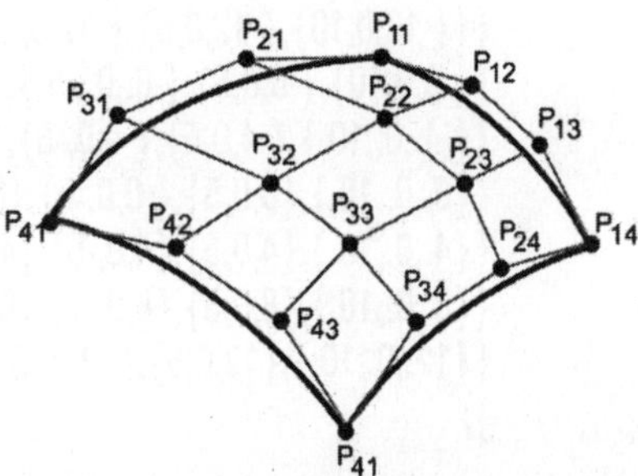

Fig.7.9: Control points for a patch

The continuity constraints on surfaces are the same as those for curves. C0 continuity implies that the two patches meet at the boundary curve. C1 continuity implies that the two patches have a smooth transition at the boundary curve, etc.

The basis function used for Bezier curves can be extended along two parameter domains to define Bezier surface patches. These surfaces are not very

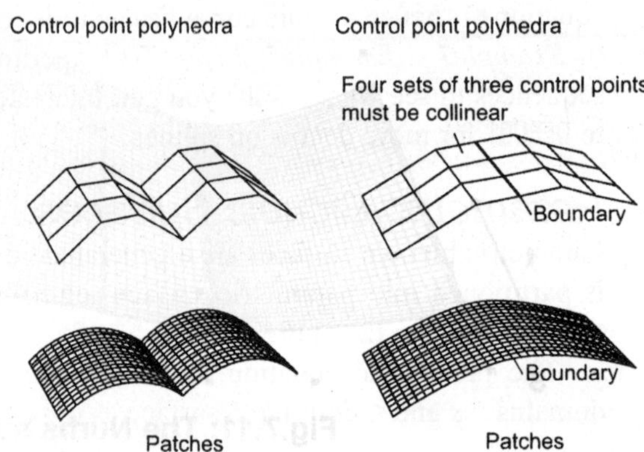

Fig.7.10: C0 and C1 continuity at surface boundaries of 2 patches

popular due to the difficulty in stitching together continuous surfaces.

Nurbs surfaces can be defined as a logical extension of the Nurbs curve equations we saw earlier. These surfaces have the same properties of Nurbs curves-C2 continuity, their ability to introduce discontinuities through knot placement and their invariance under transformations. As a result they are widely used in depicting the outer hull of surfaces.

In *Example7_3,* we define a nurbs surface and periodically move the control points to create a wave effect. We define four control points along the t domain, for each of the seven slices along the s domain. The points define a rectangular grid in the x-z plane within a boundary of (-12,12) along the x- and (-10,10) along the z-axes.

```
// s t (x,y,z)
GLfloat ctrlPoints[7][4][3] =
{{{-12,0,10},{-12,0,5},{-12,0,-5},{-12,0,-10}}, // s=0
{{-8,0,10},{-8,0,5},{-8.,0.,-5.},{-8.,0,-10}}, //s=1
{{-4.,0.,10.},{-4,0,5},{-4,0,-5},{-4,0,-10}},//s=2
{{0.,0.,10.},{0,0,5},{-0,0,-5},{0,0,-10}},//s=3
{{4.,0.,10.},{4,0,5},{4,0,-5},{4,0,-10}},//s=4
{{8.,0.,10.},{8,0,5},{8,0,-5},{8,0,-10}},//s=5
{{12.,0.,10.},{12,0,5},{12,0,-5},{12,0,-10}}//s=6
};
```

The knots along s and t are defined as

```
GLfloat Knotst[8] = {0.,0.,1.,1.,2.,2.,3.,3.};
GLfloat Knotss[11] = {0,0,0,0,1.,2.,3.,4.,4.,4.,4.};
```

The knot sequence defines a surface that interpolates the first and last set of control points along the s domain, as shown in the Fig.7.11.

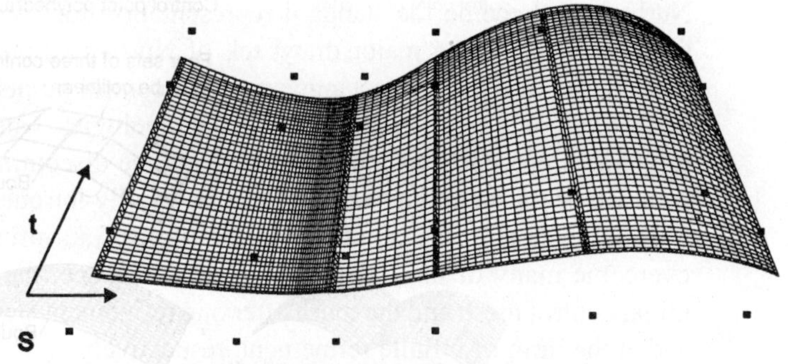

Fig.7.11: The Nurbs surface

The surface is defined by the OpenGL call

```
gluNurbsSurface.
```

This function has parameters similar to the curve but needs parameters against two domains, s and t.

```
gluBeginSurface(pNurb);
gluNurbsSurface(pNurb,
11,Knotss,
8,Knotst,
4*3,                    //stride along s
3,                      //stride along t
&ctrlPoints[0][0][0],
4,                      //order along s domain
4,                      //order along t domain
GL_MAP2_VERTEX_3);
gluEndSurface(pNurb);
```

We can further enable lighting for the Nurbs surface, by defining lights and material settings in the scene. We define our Nurbs surface to have a blueish material and set up some default lights. In order to render the surface, we need to enable automatic normal vector calculations for the surface by calling:

```
glEnable(GL_AUTO_NORMAL);
```

The complete code is given in *Example7_3/Example7_3.cpp*. We have also put in a timer that moves the control points to simulate a wave animation on the surface.

Subdivision Surfaces

Nurbs have become the standard representation for complex smoothly varying surfaces. However, a major drawback of Nurbs is the requirement that control nets consist of a regular rectangular grid of control points. Subsequently, a Nurbs surface can only represent limited surface topologies. Nurbs also they lack a local refinement procedure of the model and surface discontinuities (places where we want C0 or C1 continuity only) cannot be locally introduced or controlled.

Subdivision surfaces-defined as the limit of an infinite refinement process-overcome many of these deficiencies. For instance, the images above show an initial control mesh and the mesh after one refinement step, after two refinements, and at the limit of infinite refinement, respectively.

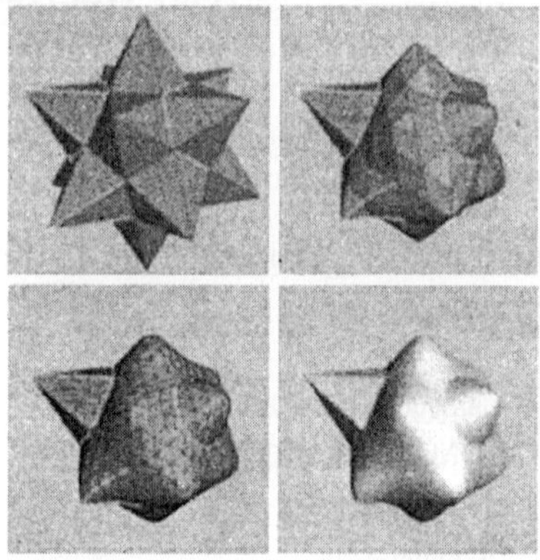

Fig.7.12: Subdividing a surface

Subdivision surfaces are easy to implement. They can model surfaces of arbitrary topological type, and as shown above, the continuity of the surface can be controlled locally. They are becoming the de-facto standard in the CG industry to model complex surfaces. Refer to DERO98 for more information on subdivision surfaces.

CSG Models

CSG attempts to build objects by adding and subtracting basic shapes to generate more complex shapes. For example if we take a cube and subtract a sphere from it, we could define a bathroom sink model as shown

Many more modeling techniques exist. Particle systems are used heavily to model particle-based objects like fire or water. Another commonly used technique to get the model into the computer is to scan it in using a scanner. The scanner digitizes the important features of the model and records these as data points. These points can then be used to generate component polygons or be approximated into a Nurbs surface. Each technique is useful in its own way and has its own drawbacks. You should pick the technique most suited to your situation at hand.

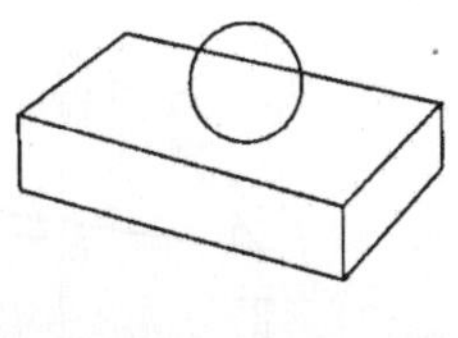

Fig.7.13: CSG to define a sink

And now, let's look a little bit more into some advanced techniques to render photorealistic images.

7.2 Advanced Rendering Techniques

Shading techniques such as those discussed in Chapter 6 go a long way towards creating colorful images. But they fall short of achieving photorealism. Real objects cast shadows (shadowing), some of them reflect other objects (reflection), and some of them allow light to pass partially or wholly through them (transmission). There are ways to "cheat" such effects. We have seen a saw a way to use texture mapping to render objects reflecting their environment. There are even techniques to fake shadows, but none of them can take into account all the various lighting effects found in the real world.

The interplay between various lighting effects is complex. One of the best ways to take all the different lighting phenomena into account is by a technique called ray tracing. Ray tracing can create stunningly photorealistic images (see Color Plate 5). Its greatest drawback is that it is very slow and mathematically very complex. It is usually combined with other rendering techniques to achieve the desired effects.

Ray Tracing

Conceptually, ray tracing is a simple process. We simply need to trace the path of all the light rays in the scene, bouncing them around along the path would follow in real life (based on the physics of light we learned about earlier). We can then combine the color of all the rays which strike our image plane to arrive at the final color for each pixel. (Contrast this approach with that used in Chapter 6, where we only used rays emanating from a light source that got reflected and directly bounced into the eye.) Fig. 7.14shows a scene of a room with a table, a martini glass, a mirror, and two sets of lamps. Rays from the lights bounce around the room to illuminate it.

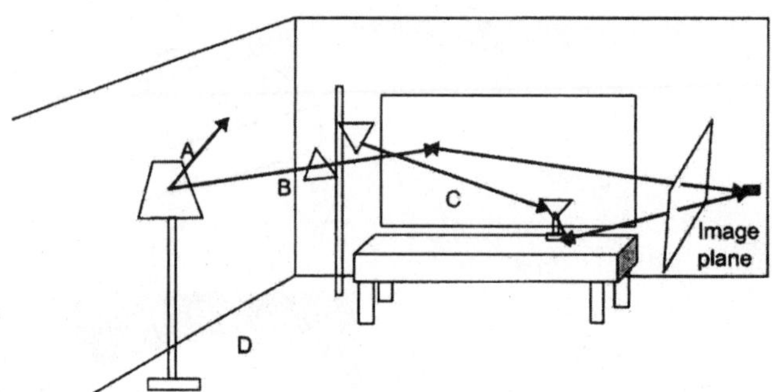

Fig.7.14: Light rays bouncing in a room

Let us trace the path that some of these rays would actually follow in real life. Ray A from the lampshade, strikes the wall and is absorbed by the wall. Its contribution to the scene ends here. Ray B hits the mirror and is reflected into the image plane to finally enter the eye. We should see the image of the lamp at this pixel. Ray C is transmitted through the glass, reflects from the table, and is again reflected into the image plane. We should see the table color with the color of the reflected image of the glass at this pixel. In general, the light rays leave the light source and bounce around the scene. The various reflections and refractions of the rays cause the color of the ray to change. Only the rays that hit the image plane, however contribute to the image. The color of the ray when it strikes the image plane will be the color contributed to the image. Some areas are never lit because rays from the light are obstructed by another object, causing shadows. Color Plate 5 shows a ray traced image of such a scene generated using Maya. We have just been ray tracing; we have followed the path of the rays as they bounce around the scene. More specifically, we have been forward ray tracing, tracing the path of light starting from their origin. Forward ray tracing is very inefficient, as most of the rays do not actually contribute to the image.

Tracing backward from the viewpoint to the light sources increases the efficiency of the process by making sure that we only trace the paths of those rays that finally reach the eye. In fact, ray tracing always refers to backward ray tracing.

The process of backward ray tracing works as shown in Fig.7.15. For every pixel of the image plane, a ray is drawn from the eye through the pixel and then intersected with models in the scene. Each time a ray strikes a surface, additional rays are drawn based on the physic of light. If the surface is reflective, a reflected ray is generated. If the surface is transmissive, a transmitted ray is generated. Some surfaces will generate both kinds of rays.

In Fig.7.15 we show one such ray striking the table. The table is a reflective surface. We reflect this ray backward, to strike the martini glass. This ray will produce the reflected image of the glass on the table surface. The glass is transmittive as well as reflective, so two kinds of rays can be spawned at this point, causing reflections in the glass as well as the "see through" appearance of

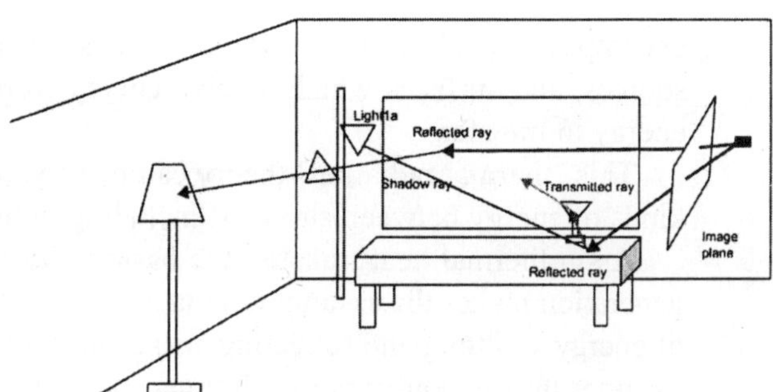

Fig.7.15: Backward ray tracing

the glass. At every point of contact, a shadow ray is drawn from the point of intersection to each light source in the scene. If the shadow ray strikes another surface before hitting the light, the surface of intersection is in the shadow of the obstructing model. In our case, the point of contact of the ray with the table is in the shadow of the second glass from Light 1a. This will cause a shadow of the second glass at this point.

Mathematically, all these rays with material characteristics of the models tell the computer about the color and intensity for every pixel in the image. After a few bounces, the rays are so weak they don't contribute much to the scene. To avoid infinitely long computations, we usually we set a limit to how "deep" we trace the rays. For example, we may set the trace to be only 3 levels deep. We would then stop tracing the rays further. We shall actually generate ray-traced images in the next chapter. Refer to [GLAS89] for more details on ray tracing and how to implement your own ray tracer.

Radiosity

Simple recursive ray tracing has many unquestionable advantages. It does a great job of modeling shadows, specular reflection, and (dispersionless) refractive transparency.

Unfortunately, ray tracing also has a number disadvantages:

- Some ray tracing methods are extremely computationally expensive and require much time to produce an image.
- The rendering time is greatly affected by the number of textures and light sources and by the complexity of environment.
- Ray tracing is view dependent, which means that the all pixel values have to be re-calculated for every new position of the observer.

To satisfy demand for generating photorealistic images, without ray tracing's imperfections, computer graphics researchers began investigating other techniques for simulating light propagation. In the mid-1980s, radiosity, as this technique is called, was finally developed.

The radiosity method of computer image generation has its basis in the field of thermal heat transfer. Heat transfer theory describes radiation as the transfer of energy from a surface when that surface has been thermally excited. This encompasses both surfaces which are basic emitters of energy, as with light sources, and surfaces which receive energy from other surfaces and thus have energy to transfer.

This "thermal radiation" theory can be used to describe the transfer of many kinds of energy between surfaces, including light energy.

As in thermal heat transfer, the basic radiosity method for computer image generation makes the assumption that surfaces are diffuse emitters and reflectors of energy, emitting and reflecting energy uniformly over their entire area. It also assumes that an equilibrium solution can be reached; that all of the energy in an environment is accounted for, through absorption and reflection.

In an enclosed environment conservation of energy assures that the energy leaving a given surface is incident upon other surfaces, thus a system of equations (the radiosity equations) may be formed describing the interdependence of the surface energies.

All that is needed to calculate these equations is the geometry of the scene, the energy of the light sources (measured in radiosity units, energy per unit area per unit time), and the diffuse reflectivity at every point on every surface in the scene.

It should be noted the basic radiosity method is viewpoint independent: the solution will be the same regardless of the viewpoint of the image and needs to be calculated only once per scene.

Like ray tracing, radiosity equations produce stunning images. Refer to [DUTR03] for the latest in radiosity technology. Radiosity is able to very effectively handle a phenomenon called color bleeding: where one surface's color seems to bleed onto neighboring surfaces. However, it operates at the cost of high memory usage and also cannot account for specular highlights since it assumes that every surface is a diffuse reflector. Probably, in the future, we shall see techniques that combine ray tracing with radiosity to achieve even more stunning effects.

Summary

In this chapter, we have learned some advanced techniques to model and render CG scenes. We have used OpenGL to define and display cubic curves and surfaces.

In the next chapter we introduce you to Maya. Using Maya, we will model a 3D scene and render it using ray tracing

Chapter 8

And Finally, Introducing Maya

You may have realized how difficult it is to visualize 3D space in your mind or on a piece of paper. You may be thinking of creating cool models, but you don't know where to position the points to get the model of your dreams! Never fear modeling software is here.

Modeling software was developed to aid in the 3D visualization process for modeling. Most 3D software now have a toolkit for rendering and animation as well.

The 3D software package we discuss in this book is called Maya. Maya has been developed by a company called Alias/Wavefront. Academy Award winning Maya® 3D is the world's most powerful 3D animation and visual effects software for the creation of advanced digital content. Maya provides a comprehensive suite of tools for your 3D content creation work, ranging from modeling to animation and rendering to name but a few.

If you have access to Maya, great. Otherwise, you can download the personal learning edition of Maya (Maya PLE, its free) from the Alias web site. Refer to the Appendix on how to download and install Maya PLE.

In this chapter, you will lean the following:

- The basics of working with Maya
- Using Maya to model polygonal and Nurbs surfaces
- Using Maya to define material properties
- Rendering a scene using ray tracing

8.1 Maya Basics

Critical to learning any software program is an initial understanding of some basic concepts: The UI of the software, the hot keys and terminology used by the software, etc. Maya is a tremendously powerful toolkit. We cannot imagine teaching you all that it can do in one chapter. However, with a few basics in place, it will be easy for you to unleash its power.

For clarity, we use a number of conventions.

When you are instructed to select a menu item we use the following convention:

- *Menu > Command (For example, File > New Scene)*

When you are instructed to select the option box for a particular menu item, we use the following convention:

- *Menu > Command > Option (For example, Create > NURBS > Primitives > Sphere >* ❑ .

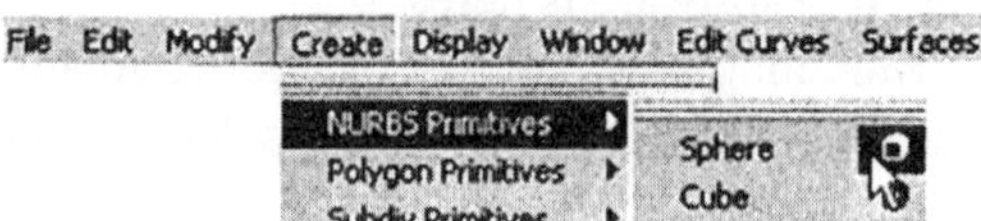

- We use the same convention for selecting from the shelf or for any other part of the Maya interface.

To start Maya

Do one of the following:

- Double-click the Maya icon on your desktop.
- (For Windows 2000 Professional) From the Windows Start menu, select Programs > Alias>Maya (version number) > Maya (Complete, Unlimited or Personal Learning Edition)

Let us first understand what we are seeing and how to use the interface. There are a lot of items displayed in the Maya user interface. We won't look into all these components right away. We will start with just the ones we need to create our first few models. As we need to use more interface components, we will introduce them to you.

The User Interface

The user interface refers to everything that the Maya user sees and operates within Maya. The menus, icons, scene views, windows, and panels comprise the user interface.

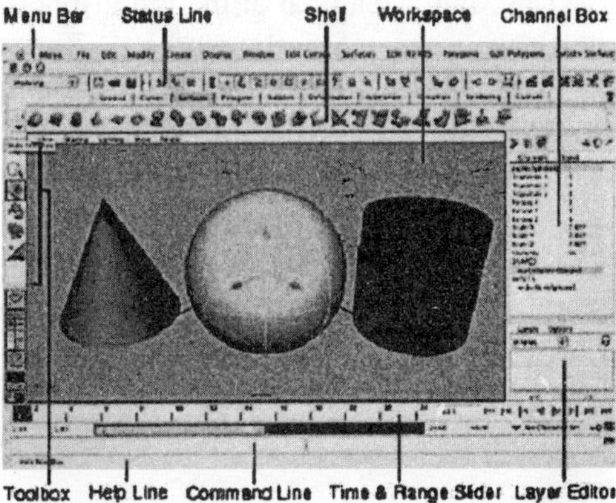

Fig. 8.1: The Maya User Interface

When you start Maya for the first time, the workspace displays a perspective projection into the 3D world. This window is called the perspective window. The window displays a grid (also called the ground) in the *xz* plane. The intersection of the two heavy lines(the *x* and *z* axis) is our origin. The y-axis points upwards. You should be familiar with these concepts by now.

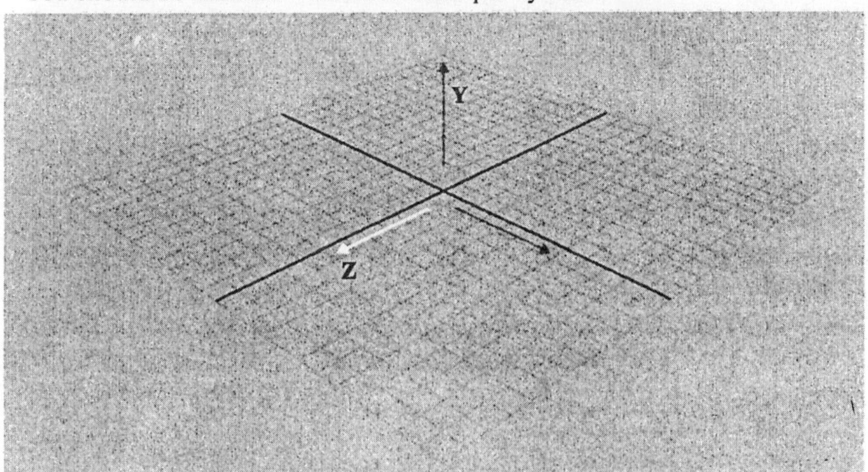

Fig. 8.2: The coordinate system in Maya

Maya labels the *x*-, *y*-, and *z*-axes with a color scheme: red for *x*, green for *y*, and blue for *z*. The color scheme is a visual aid used extensively in the modeling of objects.

The Main Menu

Maya has support for modeling, animation, rendering, and various other CG related activities. Each activity corresponds to a module within Maya. Modules are a method for grouping related features and tools of the task you are performing.

The Main Menu bar at the top displays the menu set for the module within which you are currently working. You can switch between menu sets by choosing the appropriate module from the menu selector. Let us create a primitive 3D object from the Modeling menu set.

- Choose the *Modeling* module.

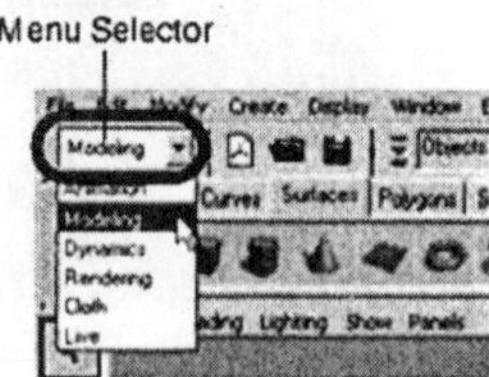

From the main menu bar

- Select *Create > Polygon Primitives > Sphere*

Maya creates a 3D sphere object and places it at the center (origin) of the Maya workspace. This primitive is defined by a polygon mesh. The model displays in a green color.

We learned earlier that a Nurbs surface (by definition) needs fewer points t han a polygonal surface to represent a smooth surface. To create the same primitive using a Nurbs surface:

- Select *Create>Nurbs Primitives>Sphere*

A sphere primitive is created which is defined by a Nurbs surface. The new model shows up highlighted in green on the screen. The previous model now turns blue.

The Status Line

The Status line is located directly below the Main Menu bar. It contains a variety of selections, most of which are used while modeling or working with objects within Maya. Many of the Status line items are represented by a graphical icon. The icons save space in the Maya interface and allow for quick access to tools used most often.

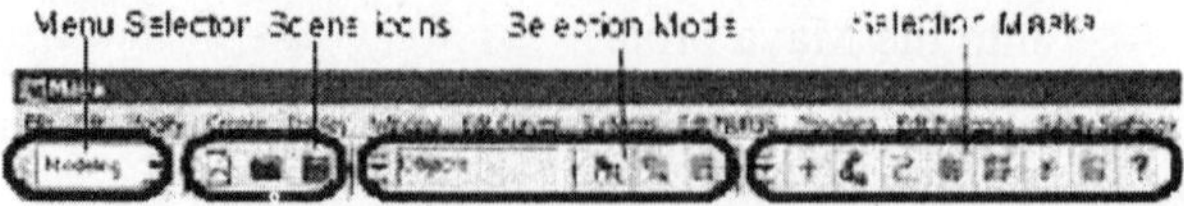

You've already learned the first item on the Status line: the Menu Selector used to select between menu sets.

The second group of circled icons relates to the scene and is used to create, open, and save your Maya scenes.

We will consider the other items shortly

Shelf

The Shelf is located directly below the Status line. The Maya Shelf is useful for storing tools and items that you use frequently use. Some of the Shelf items pre-configured for your use, tabbed into sections for convenience.

To create an object using the Shelf

- From the Shelf, select the Surfaces tab in order to view the tools located on that shelf. This tab enables you to create Nurbs based surfaces.

- Select the cylinder icon located on the left end of the shelf by clicking on it

Maya creates a cylinder primitive object as a Nurbs surface and places it at the center of the scene. By our convention, from now on, the above command will be specified as:

Select *Surfaces>Cylinder* from the Shelf.

(The same command can also be driven by the menu bar, as we saw earlier.)

In your scene, view the wire-frame outline of the spheres you created earlier. They have changed color to navy blue, and the cylinder is displayed in a bright green color. The cylinder is now the *selected* object (the sphere is no longer selected). In Maya, when the object is displayed like this, we refer to it as being selected or active.

Selection of an object or a component indicates to Maya that this particular item is to be affected by the tool or action you will subsequently choose. As you work with Maya, you will be selecting and deselecting items a lot.

To save your scene, choose the Save icon on the status line. You will be prompted to enter a name for your 3D scene. Enter a name and save your world into the Maya proprietary model file. You can always load it back up whenever you want.

You should have the basic idea now to begin!

In this chapter, we will use Maya to model and compose our 3D world and finally render the scene as a ray-traced image, as shown in ColorPlate 5.

8.2 Modeling 3D objects

We shall use Maya to model the objects shown in Fig.8.3

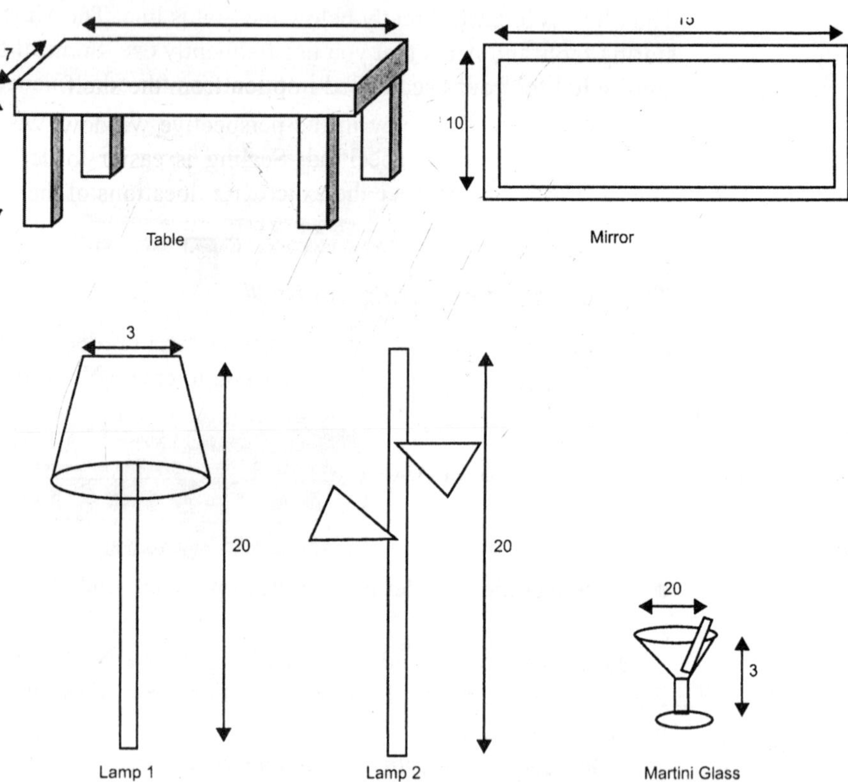

Fig.8.3: The objects to model

When you first begin to construct the models of our world, it is useful to know the size of the objects relative to each other. You can do this by drawing a rough composition of your scene and taking approximate measurements of the objects within it. This helps to avoid scaling issues later on when we compose the scene. (Of course, you can always change the sizes of the models at any time.) The basic dimensions of our objects are also shown in Fig.8.3.

The Table:

Let us start by creating the table. The table, as shown in Fig.8.3, can be thought of as a rectangular cube (the base) with thinner but longer cubes as the feet. The dimensions for this model are as follows:
15 by 2 by 7 units for the base.
1 by 6 by 1 unit for the legs.

- Start Maya, if it is not running already.
- Start a new scene by clicking the *Create New Scene* icon on the status line.

We will define the components of the table to be polygonal primitives, since we don't need the complexity of a Nurbs surface. First, we define the base of the table.

The Base

1. Choose the *Polygon>Cube* option from the shelf.

A unit cube will display in the perspective window. We want to scale this cube to the dimensions specified. Scaling is easier to define in orthographic windows, where we can see the exact *x,y,z* locations of the models.

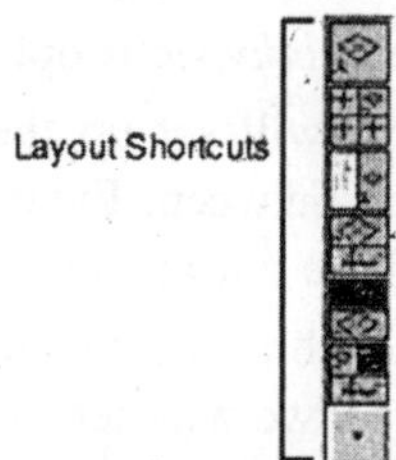

The tool-box is a UI component located on the left hand side of the Maya

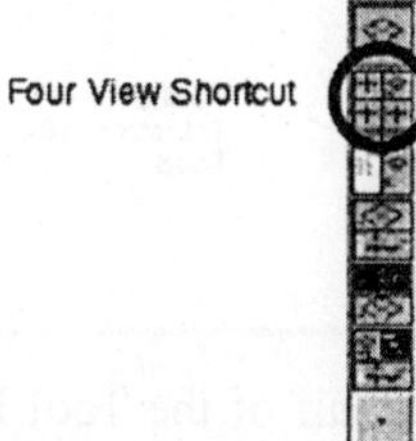

user interface. The Quick Layout buttons on the lower half of the tool-box provide a shortcut for you to select a different window panel or switch to another layout.

2. From the tool-box, click the Four View layout shortcut.

The workspace changes to a four-view layout. The three new windows define the three orthographic projection views along the *x*-, *y*- and *z*-axis (See Chapter 5).

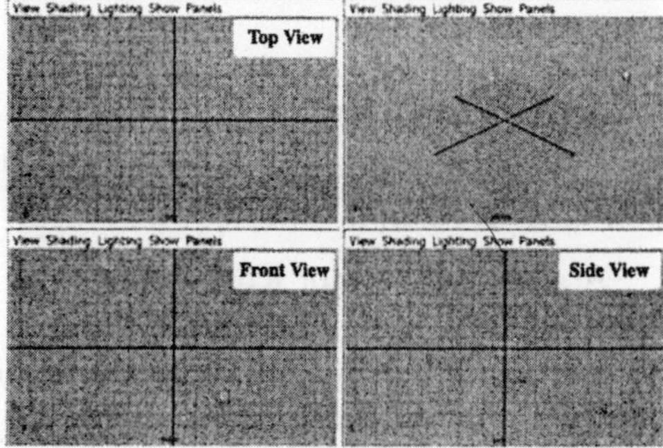

Fig.8.4: The four views: top, front, side and perpective

They are referred to as the top view, side view and front view windows. These windows are an invaluable tool in visualizing and designing the object from different angles. The perspective view is located in the top right corner, and the other views show the object from the top, front and side.

Each view has a grid defining the grid line markings along the axis. These lines are like the lines of a ruler and are an invaluable tool in determining the size of models. Let us set the markings so we can size our models within the same units.

3. Choose *Display>Grid>* ❑ from the menu bar
4. Set the *Grid Lines Every* option to be every 1 unit.
5. Set the *Subdivisions* options also to 1.

These steps will set up the grid lines to be drawn every 1 unit, with no subdivisions in between. Every grid line corresponds to one unit in our world. We will use these lines to size our models appropriately.

To size the table's width and length, we need to size it along the x-,y- and z-axes. To do this, we will need to use the Scale transformation tool located in the Tool Box.

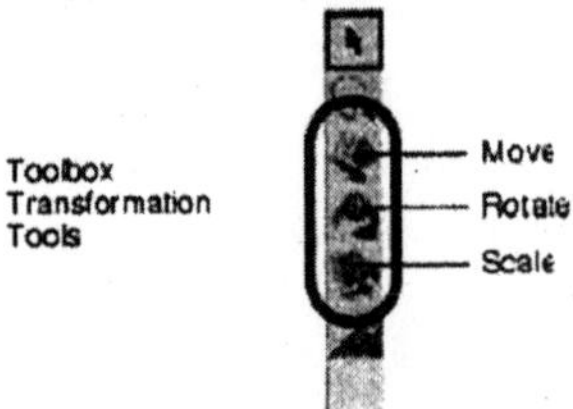

The upper half of the Tool Box contains the tools for transforming objects (selection, move, rotate, scale) within Maya. (When you move your mouse cursor over any icon, the name of the tool appears next to the mouse cursor, indicating what action you are going to commit).

Before you can transform an object, you must ensure it is selected. You can select an object by clicking on it directly, or by dragging a rectangular bounding box around some portion of the object to indicate what you want selected. To deselect an object, you simply click somewhere off of the selected object.

6. If it isn't already selected, then click on the cube with the left mouse button in order to select it.
7. Click on the *Scale* tool from the toolbox.

A scale manipulator icon appears over the primitive cylinder in the scene view.

The Scale Tool Manipulator has handles that point in the direction of the three axes. They are colored red, green, and blue, and control the direction of the scale along the axes.

When you click a specific handle, it indicates that the move is constrained to that particular axis direction.

8. In the top view window, drag the red x manipulator handle to scale the object along the *x*-axis. As you drag the manipulator, you will see the outline of the shape scaling up. Scale it enough so that it occupies 15 boxes along

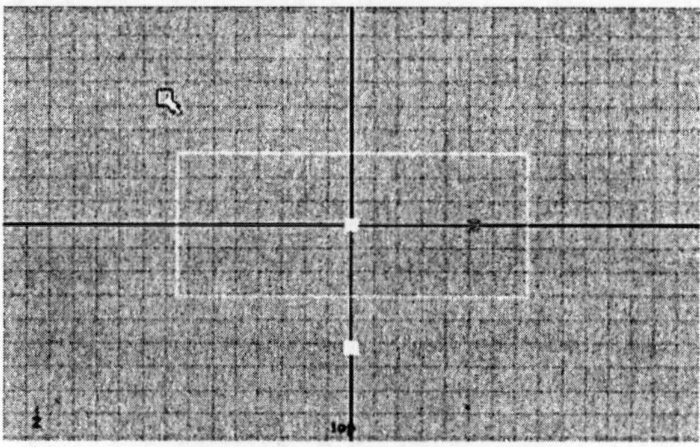

Fig.8.5: Scaling the table

the *x*-axis. This sets the length of the base to be 15 units.

9. In the top view window, drag the yellow *z* manipulator handle to scale the object along the z-axis. Scale it so that the object occupies 7 boxes along the *z*-direction, making it 7 units long.
10. To change the height of the base, click on the green y manipulator handle in the side view window, and drag it so the object occupies 2 boxes along the *y*-axis.

There! We are done creating the base of the table.
But wait! We still need to create the legs.

The Legs

The legs of the table can be defined similar to the base.

1. Choose the *Polygon>Cube* option from the shelf.
2. Scale this new cube by Sx=1, Sy = 6 Sz = 1 or (1,6,1). (Remember to select the new cube before applying the scaling transform.)

Now we need to move this leg to its correct position.

3. Select the Move Tool from the Toolbox. A move manipulator icon appears over the cube.

The Move Tool Manipulator has handles that control the direction of the movement along an axis.
When you click a specific handle, it indicates that the move is constrained to that particular axis direction.

4. In the front view window, drag the green *y* manipulator handle to move the leg downwards in the y direction. Move it downwards enough so that

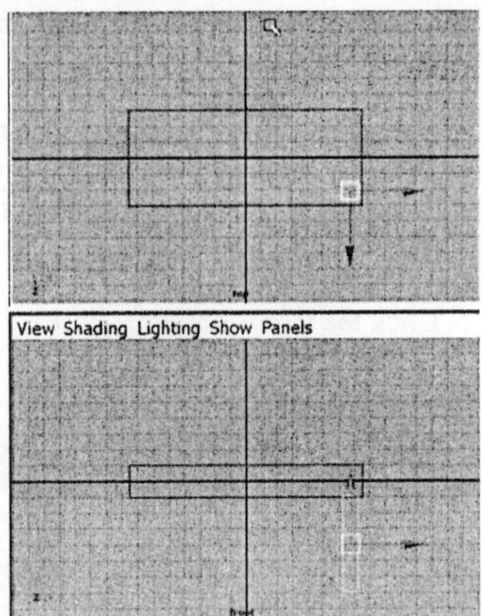

Fig.8.6: Attaching legs to the table

the top of the leg touches the bottom of the base as shown in Fig.8.6

5. Drag the red *x* manipulator and finally the blue *z* manipulator in the top view window in order to move the leg to the right hand (front) corner of the table.

We need three more legs. We can create them by duplicating the leg we already created. The shortcut to duplicating an object is the hot key: Ctrl-D (i.e., pressing the Ctrl and the D buttons of the keyboard together)

6. Select the leg. Hit Ctrl-D. This will create a copy of the selected object.
7. Choose the *Move* transform and move the newly created leg to the left corner. Repeat the process for the back two legs.

All Together now

We have a table! We should group the components of the table under a hierarchy, so we can transform the table as a single model. To do this, we perform the following steps

1. Simultaneously select all the objects that comprise the table by doing one of the following:
   - With your left mouse button, shift-click each object in turn until all the objects are selected in the scene view (that is, click on each object while holding the shift key down).
   - With your left mouse button, drag one large bounding box around the five objects using an orthographic view.
2. From the main menu bar, select *Edit* > *Group* >❑ .

This step will popup a new window called the grouping window. The grouping window allows you to group the selected objects under a new parent node (the Null Node we saw in Chapter 5). The pivot point for the parent node can be placed at either the world origin (0, 0, 0) or at the center of the grouped object. We will set it to be at the center of the grouped object. In the grouping window:

3.Set the *pivot point* option to be Center.
4. Set the *Group Under* option to be Parent.
5. Click the *Group* button.

The components will be grouped together as leaf nodes under one parent.

Maya offers different selection modes: Hierarchy, Object ,and Component. You use these modes in order to select only the types of items you want. The icons for the three modes appear on the Status Line.

When you first start Maya, the default selection mode is set to Objects. This means we select individually defined components. Now that we have grouped our components under a hierarchy, we want to select the entire hierarchy. To do this

6. Change the selection mode to *Hierarchy* by clicking on the Hierarchy selection mode on the Status line.
7. Now if you click on any component of the table, the entire hierarchy of t he table model is picked.
8. Save your work by choosing the *Save Scene* icon from the status line.
9. Save your model in a file called *Table*. Maya saves the model in its proprietary format.

Whew. One model down, four to go. By now you must have started to appreciate the power of modeling software such as Maya. But the fun is yet to begin!

The Mirror

The steps to create the mirror are similar to those we followed when modeling the table. The mirror can be modeled using two cubes with the dimensions shown in Fig.8.3. The reflective (mirror) should be slightly in front of the frame.

1. If Maya is not running, start it again.
2. Start a new scene by selecting the *Create New Scene* icon on the status line.
3. Create a polygonal cube.

We wish to scale the cube so that it has the dimensions (15,10,1). You can use the scale manipulator we saw when creating the table rr we can use the *Channels* editor. The Channels editor shows up on the right-hand side of the Maya interface and displays transformation information for the *currently* selected object.

If you don't see the Channel editor click on the Show/Hide editor for the Channel box on the status line.

4. Make sure your cube is selected.
5. In the Channels editor, set the value of ScaleX (Sx in our terminology) to be 15, ScaleY=10, and ScaleZ= 1.
6. Create another cube.
7. Set this cube to have ScaleX = 13 and ScaleY = 9.
8. Translate the newly created cube to be slightly ahead of the other cube by setting the value of TranslateZ (Tz) to be 0.3.
9. Group the objects by selecting them both and hitting the hot key to group objects--Ctrl-G -a shortcut. (If you had restarted Maya from the last model exercise, you may have to reset the grouping options by using *Edit >Group>*❑)

At this point, let us see how we can find out about the items in the scene.

10. Choose *Window > Outliner* from the menu bar.

The Outliner window pops up. The window displays a list of the items in the

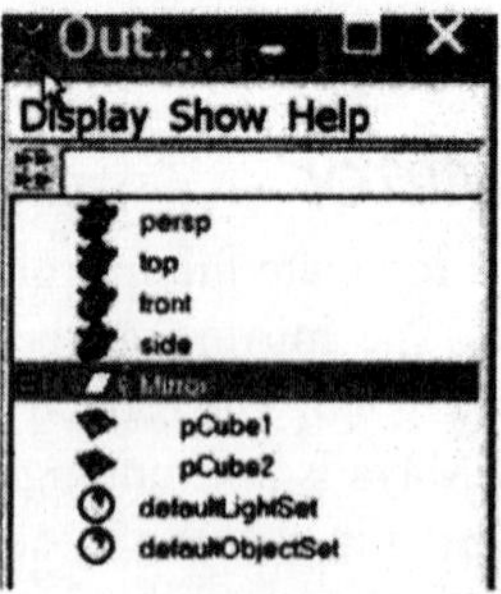

scene: including the windows and lights being used. It is very useful for grouping, setting names for models, choosing models in the scene, etc.

Your two cubes will show up as pCube1 and pCube2 (p is for polygonal), or something similar.

11. You can rename the newly created parent as *Mirror* by double clicking on the name of the group and entering the desired name.
12. Save your scene in a file called *Mirror*.

The Martini Glass

Many objects in our world have a perfect symmetry about an axis. Objects such as a vase or a glass are typical models that display such a symmetry. These kind of objects can be easily constructed by revolving a master contour about the axis

Fig.8.7: A model built by revolving a master contour about an axis.

of symmetry, as shown in Fig.8.7.

Our martini glass is a perfect candidate for a surface of revolution. Its axis of symmetry is along the *y*-axis. Let us see how to create this model using Maya.

1. Start a new scene in Maya.
2. Click on the Four View shortcut.
3. To enlarge the front view, click the mouse cursor inside the front-view window and tap the spacebar of your keyboard.

The workspace changes to a single-view layout with the front view in an enlarged view.

4. Choose *Display>Grid* ❑from the menu bar.
5. Set the *GridLines Every* to be 1 unit and the *subdivisions* to be 5. This means we have five sub divisional lines drawn for every 1 unit.
6. Since the glass has a maximum extents of only 3 units, zoom into the window by pressing the Alt key and dragging the mouse with the right mouse button held down. Zoom in so you can see around 20 grid lines.

To draw the curve defining the martini glass contour, we will define a Nurbs curve.

7. To define a Nurbs curve, choose *Create > CV Curve Tool* from the menu bar.
8. In the front-view window, click your mouse to define vertices at the numbered positions as shown in the Fig.8.8.

This will define the control vertices for our curve. Make sure the first and last positions (Control point 1 and 14) are located on the grid's *y*-axis (the thickest vertical line of the front view's grid). The first CV (Point 1) is shown as a square and the u icon depicts the tangent of the curve from the starting point.

9. Click two times in the same spot for positions 12 and 13. This action leads to the double-knot situation we saw in Chapter 7 and is useful when creating sharper corners. Three CVs at the same point would create a linear corner (three knots).
10. Press *Enter* to complete the curve creation.

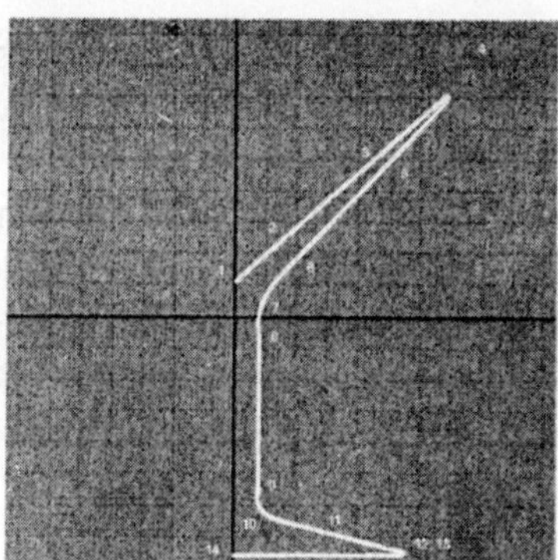

Fig.8.8: The master contour of the martini glass

If you find you need to edit the curve follow the steps below

1. Select the curve.

2. Click down on the right mouse button until a new window called the *Marking Menu* pops up.

The marking menu is used for context- related actions. The way to use it is to click the right mouse down till the menu shows up, and then drag the mouse to select the item desired.

3. Do not release the mouse. Drag the mouse to the menu item marked Control vertices to highlight this item.

4. Release the mouse button. You will now be able to see the CVs of your curve.

5. If you select the Move Tool and then click on a CV, you will see the move manipulator. Use the manipulator to move the CV around as desired.

6. Repeat step *5* until all the CVs are positioned appropriately.

11. Using the marking menu, select the Object Mode item. The CVs disappear.
12. Switch back to the four-window layout by clicking on the Four View layout icon in the toolbox.
13. Select the curve by clicking on it.
14. To revolve this curve, select *Surfaces>Revolve* from the menu bar. You will see the revolved surface appear in the perspective window.
15. Click in the perspective window.
16. You can zoom in by pressing the Alt key and dragging the mouse with the right-mouse button held down. Dragging the left-hand mouse button with the Alt key held down will enable you to view the glass from different angles. Pressing the Alt key and dragging with the middle mouse button held down will enable you to move the scene up and down.
17. Click in the perspective view, press the keyboard key *5* (a shortcut for *Shading>Smooth Shade All* option in the menu bar in the perspective window).

This displays the martini glass as a shaded surface. Spin the view around so you can view your model from all angles. Select the keyboard key *4* to display the model in wire-frame mode again.

18. If you are not satisfied with your model, choose *Display>Outliner*. The outliner window will enable you to select just the Nurbs curve you had first created.
19. Follow steps *1* to *6* to modify this curve.
20. Your surface of revolution will change its shape as you move the CVs around.
21. Save your scene into a file called *MartiniGlass.*

Stirrer

Do not create a new scene. We will create a stirrer with the martini glass as our guide.

1. Select the martini glass (and the curve used to create it).
2. Select *Display>Object Display>Template* from the menu bar. This will template the martini glass. A templated model still displays the wire-frame hull of the mode, but you will not be able to easily select it.

We wish to create the stirrer as a cylinder with radius=0.05 and height = 2. You can use the *Polygon>Cylinder* option from the Shelf as w had done earlier.

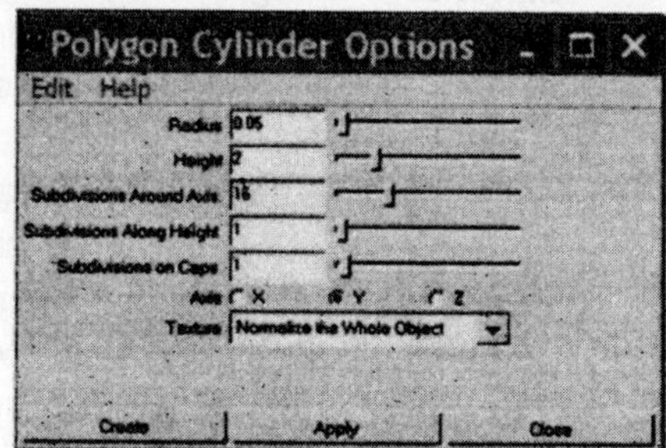

3. An easier way is to choose *Create>Polygon Primitives>Cylinder>* from the menu bar.
4. A popup a window will allow you to set the radius and length of the cylinder.
5. Set radius = 0.05 and height =2. Then click on the Create button. This will create the cylinder with the specified dimensions.

Now, we will put a sphere on the the tip of the stirrer.

6. Choose *Create>Polygon Primitives>Sphere>*❑ and set the radius to be 0.075.
7. Click Create to create the sphere.
8. Move the sphere to the tip of the stirrer by setting its TranslateY = 1.9.

Let us also create an olive on the stirrer.

9. Create a sphere with radius = 0.3. We wish to scale it along x and z to give it an elongated shape.

10. In the Channels editor (with the olive picked) set its ScaleX and ScaleZ to be 0.7.
11. Group the three objects (the stirrer, the olive, and the tip) under one parent.
12. Set the selection mode to be hierarchy and select the stirrer.
   Let us rotate the stirrer so that it resides appropriately in the glass.
13. With the stirrer selected, choose the Rotate tool from the tool-box.
   A rotate manipulator icon appears in the scene view.

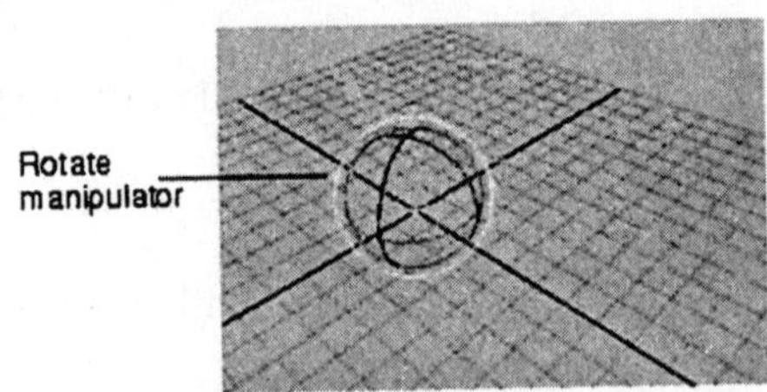

The Rotate Tool Manipulator consists of three rings (handles), plus a virtual sphere enclosed by the rings. The handles are colored red, green, and blue based on their function related to the *x*-, *y*- and *z*- axes and control the direction of the rotation around an axis.

14. In the front-view window, drag the blue Z manipulator ring to rotate the stirrer so it is aligned with the edge of the glass.

You are rotating the cylinder around its z-axis. What is the pivot point? Alternately, you can also set the RotateZ (Rz) value for the group to be about 6.3 in the Channels editor.

15. You may have to translate the stirrer further to place it within the glass.
16. To define the drink inside the glass, create a polygonal cone primitive with a radius of 1 unit and a height of 1 unit also.
17. Rotate the cone by setting its RotateZ = 180
18. Position it within the glass by settings its TranslateY = 0.7.
19. Click *Edit>Select* All from the menu bar.
20. Click *Display>ObjectDisplay>Untemplate* from the menu bar to un-template the glass.
21. View your creation in the perspective window by hitting 5 to see the objects as solid.
22. Choose *Window > Outliner.*

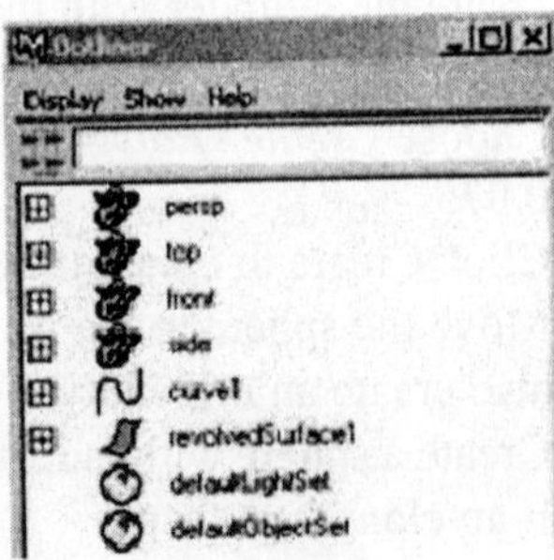

The Outliner appears and displays a list of the items in the scene.

23. In the Outliner, select the curve you revolved and delete it, so you will not accidentally modify it later on.
24. Group all the objects under one parent.
25. Rename the group to martiniglass.
26. Save your work back to the file called MartiniGlass.

Lamp1

Lamp1 has a cylindrical stand with dimensions as shown in Fig.8.3.

1. Start a new scene in Maya.
2. Reset the grid so the gridlines are drawn only once every unit.
3. Create the stand by choosing *Create>Polygon Primitive>Cylinder>*❑
4. Set the *radius* to be 0.3 and the *height* to be 20. Click Create.

The shade can be modeled as a cylinder, that has been scaled on the top to define a conical look. We will define it to be a Nurbs surface to get a smoother look.

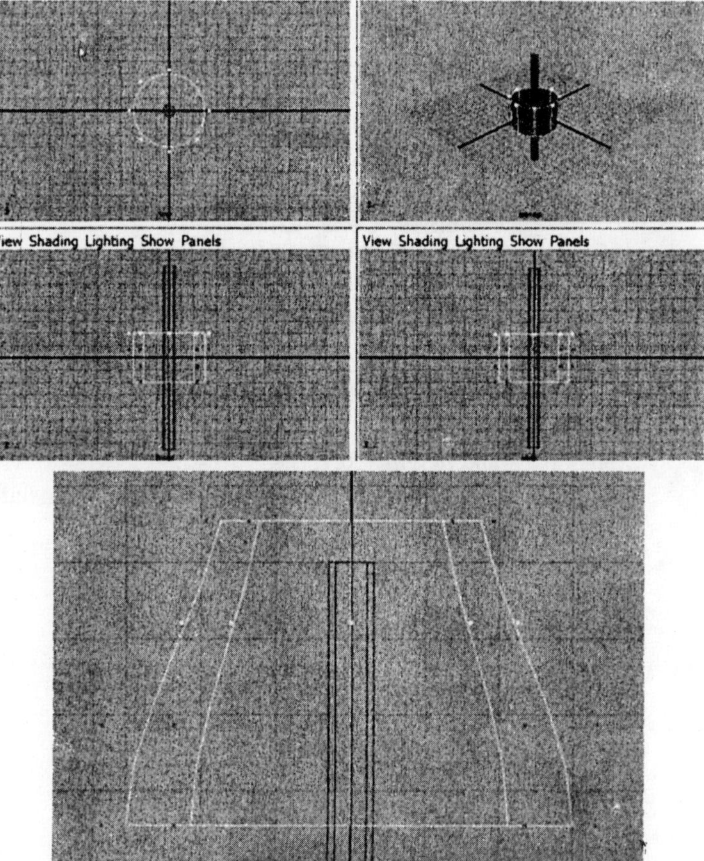

Fig.8.9: Modeling the lampshade

5. Choose *Create>Nurbs Primitive>Cylinder>* ❑ and set the radius = 3. and *height* = 4.
6. Select the cylinder and select *Control vertex* (CV)from the marking menu. You will see and be able to pick the CVs associated with this Nurbs surface
7. In the front-view window, select the top ring of CVs by dragging a box around them.
8. Select the *Scale* tool from the tool box.
9. In the top-view window, scale the ring of vertices down along *x* and *z*.
10. Do the same for the second ring of CVs, but scale it less than the first ring.

We are munging the vertices defining our Nurbs surface. Recall that this affects surface patches local to the vertices. The front view of the lamp shade should now appear as shown in Figure above.

11. View your model in the perspective window.
12. Group and name your object lamp1.
13. Save you work in a file called *Lamp1*.

Lamp2

1. Start a new scene
2. Create the stand for lamp2 as a polygonal cylinder with *height* = 20 and *radius* =0.3
3. The lampshades will be Nurbs cones. Select the *Create>Nurbs Primitive>Cone>*❑.
4. Set *radius* = 1.5 and *height* = 3. Make sure the *Caps* option is set to None.

We will create a cap for the cone, but we wish it to have a different material than the shade. This will serve as the light beam shining from inside the shade. For this, we will cap the cone with a polygonal cylinder.

5. Select *Create>Polygon Primitives>Cylinder>*❑ and set *radius* =1.3 and *height* = 0.1.
6. Translate the cylinder by settings its TranslateY= 0.25, so that it caps

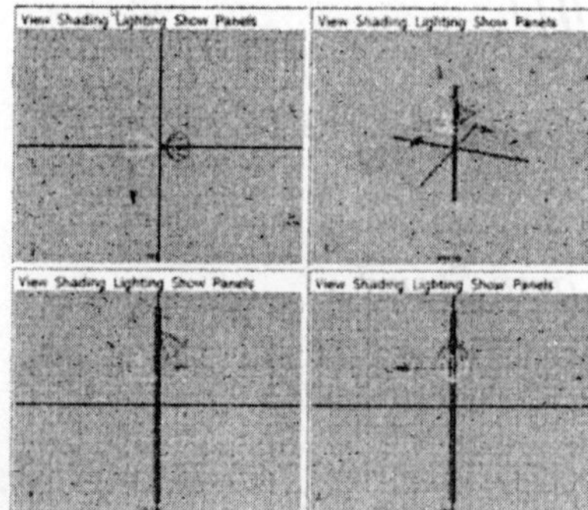

Fig.8.10: Modeling the second lampshade

the cone.

7. Group the cone and the cylinder, and call the group *Shade*.
8. Make sure the selection mode is set to *hierarchy*.
9. Select the shade and rotate it by setting RotateZ = 40.
10. Move and translate the shade by settling TranlsateX = 2 and TranslateY = 3.1
11. Lamp2 has two shades. Duplicate the shade created by hitting Ctrl-D.
12. Set the new shade to have its transforms defined as RotateZ = -120, TranslateX = -2.5, TranslateY = 1.8, so that it is on the other side of the lamp.
13. Group the three objects under one parent called lamp2.
14. Save the scene as *Lamp2*.

8.3 Applying Surface Materials

Let us apply surface materials to our objects, so that they start to look more appealing. To ensure the steps work as described, select *Rendering* from the menu selector. Tab to the Rendering tab on the Shelf. This will enable the shortcuts to the common rendering tools.

We shall start by defining the materials of the two lamps first, since these are the simplest. We will then define the material of the martini glass which needs transparency defined for the glass surface; the mirror, which needs reflectivity settings; and finally the table, which needs a texture setting for its wood.

By default, Maya assigns a default material called *lambert* to all models created. This default material is grey in color and has diffuse-only reflection defined. To assign new materials to objects, we need to create new material attributes.

Lamp1

1. In Maya, click on the *Open a scene* icon from the Status line.
2. Type in the name of the model file where you saved Lamp1. This will open up your Lamp1 model.
3. Set the selection mode to *Object* and select only the lamp stand.

The Lamp Stand

Let us first define the material for lamp stand. We want the stand to be a reflective metallic material so we will need to create a new material for this object.

1. Right-click the stand to select it.

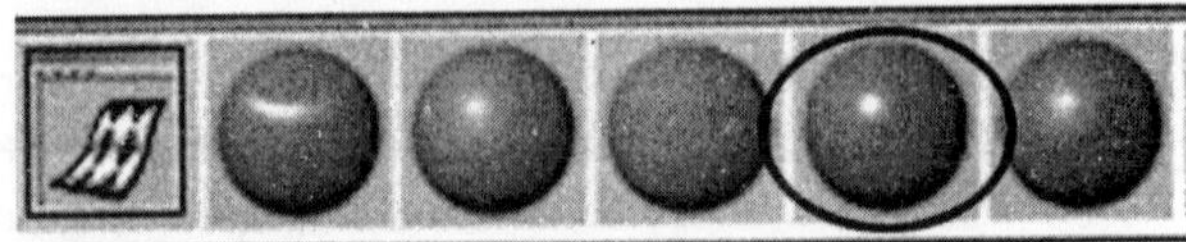

2. Select *Rendering>Assign a New Phong Material* icon from the shelf.

This option will create a new material and assign it to the lamp stand. Phong shaders includes all three components of the illumination model: ambient, diffuse, and specular components. A Phong material will give a surface a shiny look.

The Attributes editor will appear on the right hand side of the Maya UI. The editor will display a sphere rendered with the current material attributes. If you don't see the Attributes editor, click the Show/Hide attributes editors from the status line (it is to the left of the Channels editor button).

Below the sphere on the Attributes editor, is the menu to modify the attribute settings of this material.

3. Click in the text box that defines the name of the material and change it to *Chrome* so that we can identify this material for use later on in the project. The name of the material is at the topmost text box of the editor.

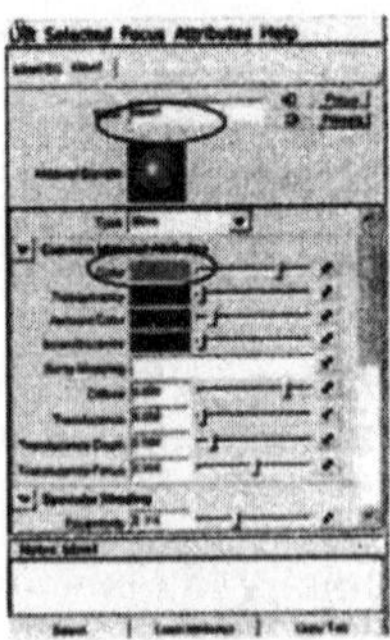

4. In the *Common Material Attributes* section click inside the color box to the right of *Color*. This will open the Color Chooser window.
5. The Color Chooser window displays a color wheel. You can click inside the color wheel (hexagon) and drag the mouse pointer to change the color of your material.

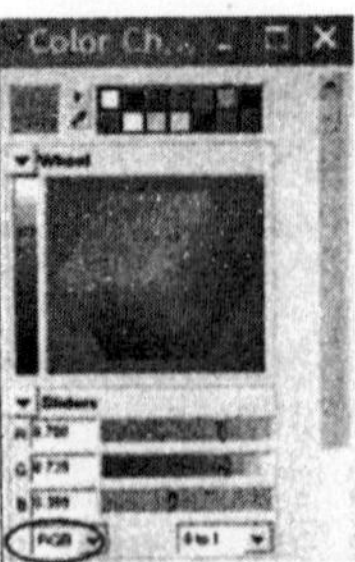

6. You can also set the color by specifying the RGB components. Change

the slider display from HSV to RGB as shown, and set the RGB values as R=0.96, G=0.96, B=0.5-a yellowish color for our chrome stand.

7. Click *Accept* to choose the new color and close the Color Chooser.
8. You can change the intensity of the diffuse (and ambient) reflectivity of your material by dragging the *Diffuse Color* (or *Ambient Color*) slider to its right. Do so to increase the color brightness.
9. You can change the *specular settings* of your material by using the Specular Shading section in the attributes editor. The Specular Color option changes the specular reflectivity of the material (and its intensity), and the Cosine Power slider changes the size of the shiny highlights.

10. Ignore all the other attributes for now. Click in the Perspective window.
11. Select the *IPR Render Current Frame* icon from the Shelf.

This will render an image of the currently selected window. Since you clicked in the perspective window earlier, this will render an image of your lamp as seen in the perspective window.

12. You can change the perspective camera (by hitting Alt and Mouse right/middle/left) to view the lamp from different directions. Hitting the IPR button will render the lamp from the new viewpoint.
13. Change your material until you are satisfied with your render.
14. If you lose the material editor, choose your lamp stand again and choose the *Show shading group attribute* editor icon from the Shelf. You can then tab to the Chrome material in the Attributes editor

The lampshade

The lampshade is made of cloth and has a diffuse material. We will create a new Lambert material for the shade. Lambert shaders have only ambient and diffuse components and hence do not look very shiny-Perfect for a cloth.

1. Select the *Rendering>Assign a new Lambert material* icon from the Shelf.

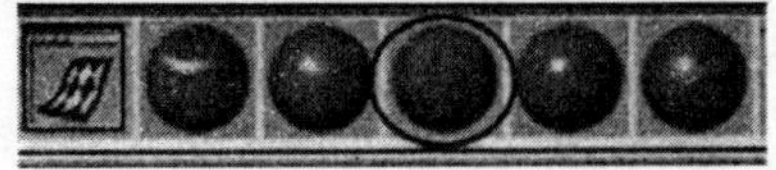

2. Set the name of this material to be *shade*, and set its color to be R=0.9, G=0.5 B=0-an orangish tinge

When you see a lamp shade in real life, it seems to glow from the light within it. We can achieve this effect by giving the shade a bit of emissive color.

3. To do this, drag the Incandescence slider, under the Common Materials *Attributes section*, to the right. The Incandescence of a material defines the emissive nature of the material.
4. Render till you are pleased with the results!
5. All done with lamp1! Save your scene back to avoid losing your work.

Lamp2

1. Select the *Open a new scene* option and open the file Lamp2.
2. Set the selection mode to be *Object* mode.
3. Select the stand and assign a new Phong material to it.
4. Set the name of the material to be *metal*.

We shall assign a metallic-looking material to the lamp and the lamp shades.

5. Assign this material to have a whitish color. RGB = (0.8, 0.8, 0.8).
6. Under the *Specular* shading options, set the specular color to be the same color.

If you have looked carefully at a metallic object such as a spoon, you may have noticed that you can see reflections in it. Metallic objects reflect their environment. Maya enables materials to reflect their environment by the value of the Reflectivity attribute. The slider to control this attribute ranges from 0 (no reflections) to 1 (clear reflections). The default value is 0.5. Reflectivity values for common surface materials are car paint (0.4), glass (0.7), mirror (1), metal (0.8).

7. Set the reflectivity value (under Specular Shading options) to have a value of 0.8.
8. You will not see the reflections in the IPR render. When we ray-trace this object, reflections in the object will give it its true metallic look.
9. Pick the cones of the shade and set them to have the metal attributes.

You can do this by selecting *Materials>Assign Existing Materials>Metal* from the marking menu.

10. Pick the cylindrical caps of the shade.

We want the lamp shade to look like it's glowing-as if the light is coming out of this object.

11. Select *Rendering>Assign a new lambert material* icon from the shelf.
12. Set the name of this material to be *shade2*, and set its color to be R=0.9, G=0.9 B=0-a yellowish tinge.
13. Drag the Incandescence slider to the right. This component will change the emissive color of the material.
14. Render.
15. Redefine the materials until you are pleased with the results!
16. Save your scene.

Martini Glass

1. Load up the martini glass.
2. Pick the revolved glass object.
3. We need to assign this object a glass material. Glass is transparent, but it still displays specular highlights and some reflectivity.
4. Assign the glass a new Phong shader and call this shader, *glass*.
5. Leave the color of the glass as is.
6. Glass is transparent: we can see through it. To set the transparency of the material, slide the Transparency slider. You can slide it to the right to get a nice transparent glassy look- but don't drag it all the way to the right, or your object will become invisible!
7. Under the Specular Shading options, we set the Reflectivity of the glass to be 0.2. This will make the glass show very faint reflections.

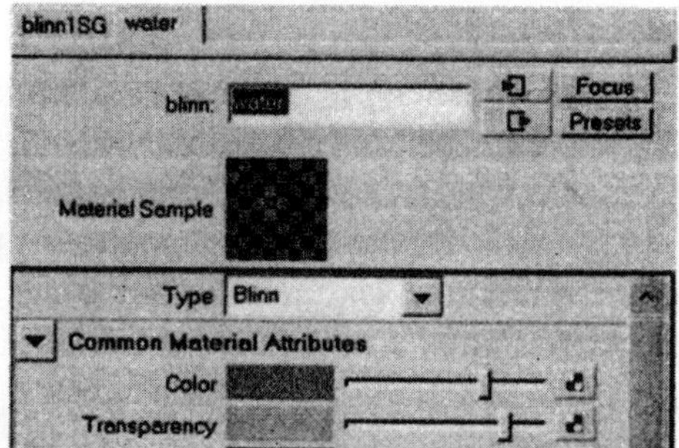

The cone inside the glass is made of water. Water is also transparent, but it is usually perceived as blue in color.

8. Pick the cone inside the glass. We will make this object be a water material.
9. Assign it a Phong shader.
10. Set the color of the water to a bright blue. RGB= (0,0.9,1)
11. Slide the Transparency slider to get a nice transparent water look.
12. Again, we made the water nonreflective, although you can define a small amount of reflectivity to it if you like.
13. Set the stirrer to also have the shader glass, by selecting *Materials>Assign Existing Material>glass* from the marking menu.
14. If you lose your material, remember. You can always pick the model and click on the *Show shading group attribute editor* icon from the Shelf.
15. Create a new green-colored Lambert shader, with RGB=(0,0.5,0) for the olive.
16. Save your scene to avoid losing your changes.

Mirror

1. Load the mirror.

2. Select the outside cube-the frame of the mirror.
3. Assign it a new lambert shader.
4. Set the color of the shader to be magenta: RGB = (0.8, 0,0.2)
5. The mirror itself has a black color, and is completely reflective.,
6. Select the mirror.
7. Assign a new Blinn shader to it, and call the material mirror.
8. Set the color of the mirror shader to be black and the diffuse color also to be black. The mirror itself dose not have any color. It merely reflects its surroundings.
9. In the *Specular Shading* section, set the specular color of the mirror to be 1 by sliding the slider all the way to the right.
10. Set the *reflectivity* of the mirror to be 1. The mirror reflects everything!
11. Save your scene

Table

1. Load the table model.
2. Pick the base of the table.
3. Assign it a new Phong material. Name the shader wood.
4. Click the checker icon next to the Color slider. This is a quick way of applying a texture.

A window called the Create Render Node window pops up and displays a list of the textures that you can apply. The texture names and their icons help you to decide which texture is appropriate for the effect you want.

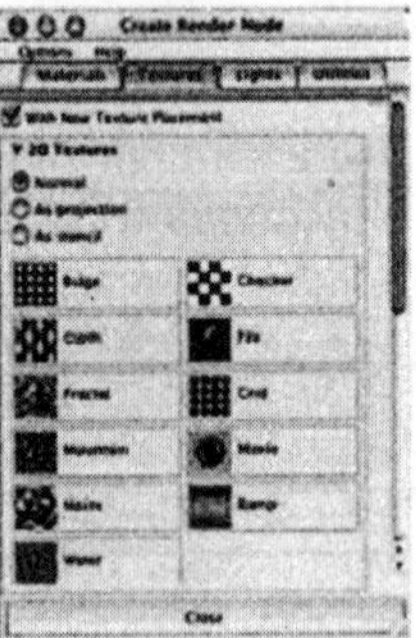

Maya has both 2D textures and 3D textures. We saw 2D textures in Chapter 7. They are essentially 2D images that you can map onto your object using a planar mapping or by shrink-wrapping it onto the surface. In Maya, a 3D texture is essentially a cube, with texture information on all sides of the cube. The texture is then mapped onto the object using the cube mapping we saw in Chapter 7, making the object appear as if it were carved from a block of the material selected. We shall assign a 3D wood texture to our table.

5. Under the 3D textures option, choose the wood texture.

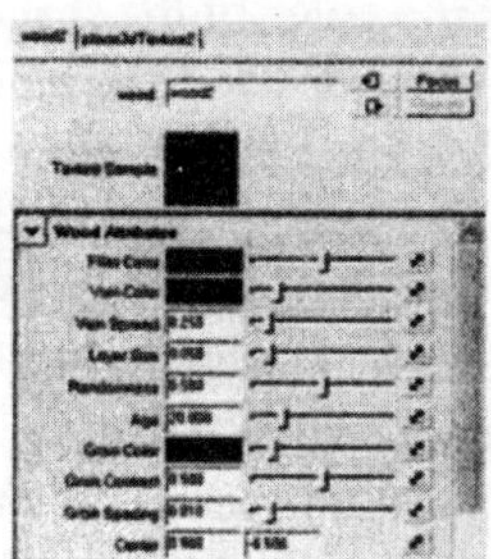

6. A green box indicating the position of the texture cube will be displayed in the windows.
7. The various parameters to define the values for this texture will appear in the attributes menu.

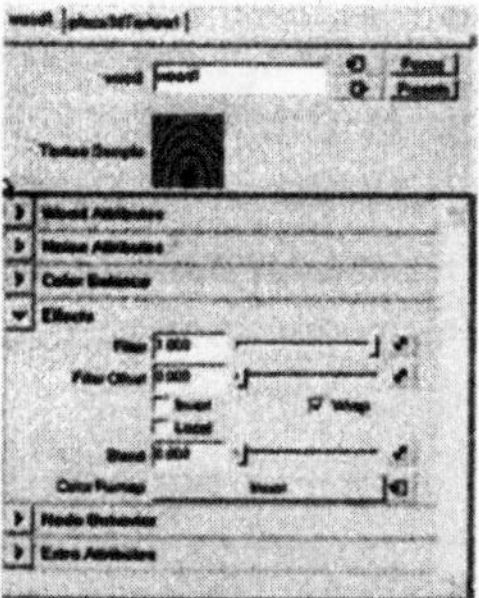

8. Under Wood Attributes, click on the filler color. In the color selector, set the RGB color of the Filler to be R=0.5, G= 0.26, B = 0.0 to get a rich brown color, which is what oak wood normally looks like.
9. Under Effects, turn on the Wrap radio button, so the texture will repeat itself. .
10. Render the table.

You will notice that the wood grain is closely packed. We need to scale the texture box up to scale out the wood grain. Also, the texture needs to be rotated so the texture cube is mapped correctly. We will need to rotate the texture box to achieve this.

11. Click on the place3DTexture1 tab in the Attributes menu.

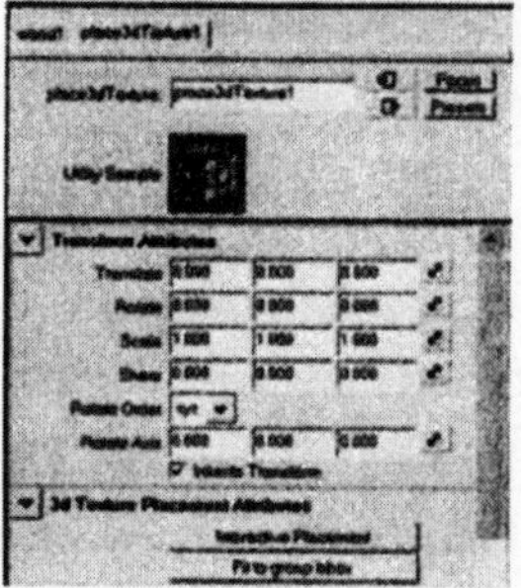

This tab will enable you to position and scale the texture appropriately. Essentially, you will be applying transforms in the texture space.

12. Set RotateX =-90 and RotateY = 90
13. Set ScaleX =2, ScaleY= 4 and ScaleZ = 1.

Rotate	-90.000	90.000	0.000
Scale	2.000	4.000	7.000

You will see the green texture box now engulfing more of the table.

14. Render.
15. You should see the tabletop rendered with a nicely defined wood texture.
16. Assign the wood texture to each of the legs by selecting each one and choosing Materials>Assign Existing Materials> Wood from the Marking menu.
    In the Attributes menu, you can navigate to the wood shader by selecting any object in the scene which has the wood shader defined. You can navigate from the wood shader to the wood texture by clicking the icon next to the Color slider.

When you render the image, the legs will not have the texture mapped onto it. This is because the texture bounding box does not engulf the legs.

17. Scale up the wood texture placement by setting ScaleZ = 7.
18. Render.
19. Save your scene.

8.4 Composing the World

Let us now put all our models into the world to compose the scene as in Color Plate.

1. If you have closed Maya, restart it. Open a new scene.
2. Load up the Table model.
3. Choose *File>Import* from the Menu.
4. Choose the file: *Lamp2*. This imports the lamp2 model into our scene.
5. Set the selection mode to be *hierarchy*.
6. Pick the lamp2 object.
7. Click on the Channels attribute box, and set the transforms for lamp2 to be
   - TranslateX = -2.7, TranslateY = 5.8, TranslateZ = -2.25
   - RotateX=0, RotateY = -15, RotateZ = 0
8. Import the martini glass.
9. The glass should rest on the base of the table. Set its translations to be
   - TranslateX = 3.59, TranslateY = 0, TranslateZ = .472

10. Duplicate the martini glass by hitting Ctrl-D
11. We want this second glass to sit beside the first glass. Set the transforms for the second martini glass to be
    - TranslateX = 0.46, TranslateY = 2.5, TranslateZ = 0.5; RotateY = 158
12. Import the mirror. It resides behind the table, hanging on a (as yet imaginary) wall. Set its transforms to be
    - TranslateX = -0.7, TranslateY = 5.437, TranslateZ= -5.125

In the perspective window, you will be able to see your world slowly being defined. You can change the perspective camera to a camera setting that pleases you. (Do this by hitting Alt and dragging the mouse with one of the buttons pressed down)

13. Import Lamp1. Lamp 1 will be positioned behind our camera, and it will only be seen as a reflection in the mirror.
14. Zoom out in the top view and position the lamp in the front left hand corner.
    - TranslateX = -10, TranslateZ = 23

We wish to define our 3D world within the confines of a 3D room. We need walls for the room that we are position in!

15. Create a new unit cube.
16. Scale it to ScaleX = 35, ScaleY=35, ScaleZ = 35.
17. Position it as shown in Fig.8.11 by setting its translations as
    - TranslateX = 2, TranslateY = 11, TranslateZ = 11.

Assign a new lambert shader to the wall with a yellowish color.

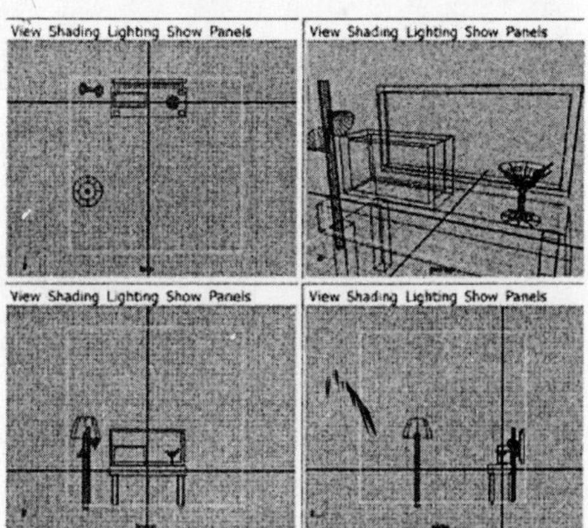

Fig.8.11: World being defined in Maya

Click in the perspective window to select it and render your scene.

Voia! A 3D world. The scene looks good - but the lights are still default lights. (If you are using the Learning Edition of Maya, then the render will have the Alias logo over it as a watermark).

We still need to define lights to define the mood of our world, and to render using ray tracing so we can see reflections and shadows. Save your scene in a file called *world*. We shall use this scene again in Chapter 11, when we animate the scene.

8.5 Lighting the scene

When you open a scene, Maya lights it with default lighting provided by an imaginary light fixture.

Once you define your own lights, Maya will render using your defined lights instead. Maya supports ambient light, directional lights (like sunlight), spotlights (useful for cone lights), and point lights, to name a few.

Let us define lighting for the current scene.

Ambient Lighting

First we define some ambient lighting so that nothing in our scene is rendered black.

1. Choose the *Rendering>Ambient Light* icon from the Shelf.

2. The attributes menu for the newly created light will show up.
3. Under the Ambient Light Attributes section, set the intensity of the light to be 0.2.

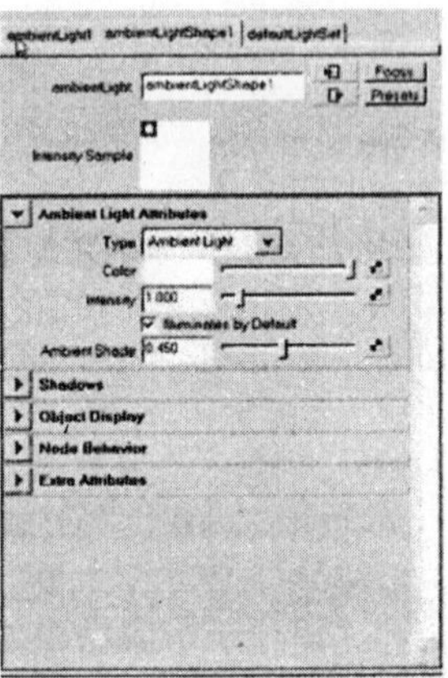

Light from Lamp1

We want to simulate lights from the lamps that we have modeled. Let us first consider lamp1. A point light is a good light to simulate light bulbs, which emit light in all directions. We shall create a point light and position it at the center of lamp1.

1. Choose the *Rendering>Point Light* icon from the shelf.
2. A point light icon will appear at the center of the grid. You can move the light just as you move objects. Translate the point light to be located where the shade of lamp1 is.
3. In the Attributes editor, set the intensity of this light to be 0.3.
4. Turn off the *Emit Specular* radio button so that this light has only a diffuse component.
5. Render your scene.

Your scene should look very moody now. We need to define the mood to be a little more upbeat by adding some more lights!

Light from Lamp2

Lamp2 can also be thought of as simulating a light bulb. However, since it has metallic shades, the light is not spread in all directions; it is restricted to a cone defined by the shade. Spotlights create a cone of light and will be ideal to define the lights of lamp2.

1. Select the spotlight icon from the rendering shelf..
2. This creates a spotlight. An icon for this light will display at the center of the grid. The light points toward the back wall.
3 If you look at the spotlight icon from several angles, you'll notice that it is shaped like a cone with an arrow pointing out of it.
4. The cone symbolizes the beam light that gradually widens with distance. The arrow symbolizes the direction of the beam.
5. With the spotlight selected, select *Modify* > *Transformation Tools* > *Show Manipulator Tool*.from the Main menu. This tool provides two manipulators that you can move to position and aim the light precisely.

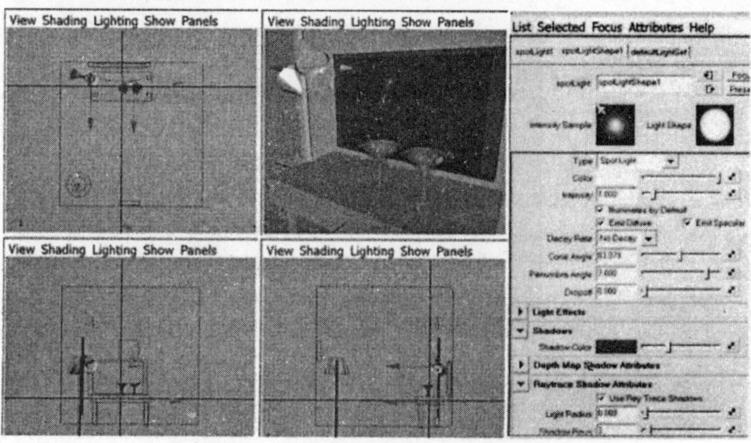

Fig.8.12: Setting up spotlights

6. The look-at point specifies where the light focuses. The eye point defines the position of the light source. All types of lights have an eye point, but not necessarily a look-at point.
7. Move the look-at point to the top of the table, near the wine glasses.
8. Move the eye point roughly to the center of the shade cylinder, as shown in Fig.8.12. (We cheated a little so that our lighting effect could be what we wanted!! Actually the light should be placed inside the shade itself.)
9. In the Attributes menu, set the color of the light to be slightly yellow in order to simulate real bulbs in housing lights. RGB = (1,1,0.7).
10. The spotlight's default Cone Angle is 40 degrees. In the Attribute editor, set the value of the *Cone Angle* to 20.
11. Render your scene. The circle of light is very narrow.
12. Set the cone angle to be about 83 and render the scene. The area of light expands and is now too big.
13. Set the cone angle to be 50 and the penumbra angle of the light to be 10. This will produce soft edges for the light.
14. In the Attributes menu, expand the Shadows section and further expand the *RayTrace Shadows Attribute* section for this light. Turn on the *Use Ray Trace Shadows* radio button. This will cause objects being lit by this light to cast shadows when you render using ray tracing.
15. Set shadow rays and ray depth limit to be 3 to ensure that shadows are created correctly.

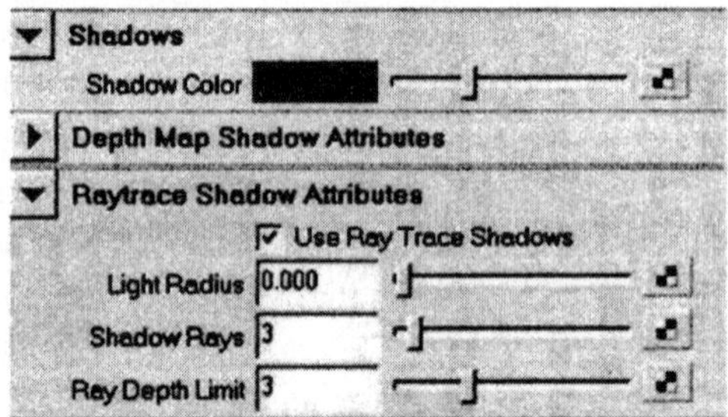

16. Usually, shadows are never stark black. They are some shade of grey, depending on the lighting. To achieve this effect, we also modified the Shadow Color attribute to be a dark grey.
17. Create a similar spotlight for the shade on the other side. You will have to position it appropriately.
18. Set this light's cone angle to be 60, penumbra angle to be 10 for a soft light on the wall, intensity to be 0.8, and color to be a light yellow. You should be able to see a cone of light on the wall when you render the image.

Render the scene - *wow*! Doesn't it look amazing? And we still haven't gotten to ray tracing. This will be the final step, when we will see reflections;

refractions and shadows all come together to enhance the visual realism of our image.

8.6 Finally – Ray Tracing

1. From the Shelf select the Display Render Globals Window icon.

This will bring up a window allowing you to set the rendering options.

2. Set the *Render Using* option to be Maya Software.
3. Under Resolution, set the resolution to 640 x 480 or higher if you wish.

This sets the rendered image to the specified size.

4. Click on the Maya software tab.
5. Under the *Anti-aliasing Quality* options, set Quality to be Production Quality-this will be a quality render.

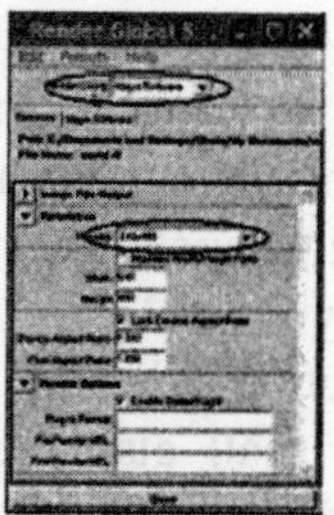

6. Under Ray Tracing quality, set Ray tracing to be on!

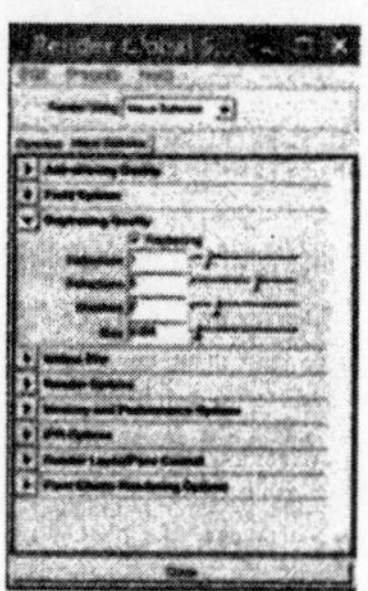

Hurray. We are going to be ray tracing our scene!
You cannot use IPR to render ray-traced images.

7. Select *Render>Render Current Frame* from the menu bar.

You will see a ray-traced image being rendered of the currently selected window. Be patient. Ray tracing can take a while. Preview your render. You may need to modify your lights, models, and shaders to get the reflections and shadows in the exact spots you want!

Only the spotlight for which we turned on the ray-trace shadows option, will be obstructed by objects in its path, causing shadows. The reason we had this light to be located outside of the shade is so that its light would not be blocked by the cylindrical cap! The other do not cast shadows, and hence, objects in their path do not obstruct the beam of light. You can play around with the light settings as you please. For the image in Color Plate 5, we actually turned on shadows for the point light in lamp1 as well.

You can also pick specific models and edit their surface materials. For example, we actually added some ambient color to the table and the glass so that they didn't render as dark. Use the Window>Outliner to select objects as desired if you have a hard time picking them otherwise. You can also find a Maya model of the world that we modeled in a file called world.mb (Maya Binary file format) under the installed directory for our software, in the folder Models.

Have fun experimenting. And yes, we did cheat and add in a texture mapped cube in our color plate. You can see it reflected in the mirror. Define your own texture image for this cube to see it in the final image.

Summary

In this chapter, we have learned how to use Maya to model and render a 3D scene. Could you visualize what code was running in the background to display the models and images? By now, you know enough to actually write a small Maya package yourself!

The next section will deal with how to make our scene come to life by adding animation. We shall also see how to add more pizzazz to our scene with additional lighting and viewing techniques.

Section 3

Making Them Move

In the last few sections, we saw how to model objects and render them to create photo-realistic images. Our next task is to give life to our creations by making them move: animation!

When you see a cartoon on TV, you see characters in motion. This motion is just an illusion: the strip itself is actually composed of a sequence of still images (or frames) which when played back in rapid succession gives the illusion of motion. The most popular way to create animation is by generating a number of successive frames differing only slightly from one another. However, just making the characters move is not enough to make an animation look believable. Walt Disney himself put forward the seven principles of animation that add zest and realism to the animation.

In Chapter 9, we shall look into the techniques and principles used to create traditional (hand drawn) 2D computer animation and how they can be adapted in a 3D world. Chapter 10 will walk you through the creation of an animated 3D game. In Chapter 11, we shall use Maya and work through the process that traditional production houses use to create our own animated production.

Hold on to your seats!

Chapter 9

Animation

To animate means to bring to life. In the last few chapters, we saw how we can use the principles of Computer Graphics to draw and render 3D worlds. Our next task is to breathe life into our models.

When you see a movie on television or in the theater, you see characters in motion. This motion is usually smooth and continuous. The actual tape containing the movie, however, consists of a sequence of images, or what are known as *frames*. These frames, when flashed in rapid succession on the screen, cause the illusion of motion. Fig.9.1 shows such a sequence of frames in which a smiley face changes into a frowning one.

Fig.9.1: A sequence of frames

The illusion of motion exists because of a peculiarity in the human vision system. When you see a picture, it tends to remain in your mind's eye for a brief period of time—a phenomenon called *persistence of vision*. If the pictures are shown quickly enough, then a new picture arrives before the old one fades out. Your eyes perceive a smooth transition between the frames. If the difference between successive frames is too large or the frames are not shown quickly enough, then the illusion shatters and the movement seems jerky. You may have seen this kind of effect in older movies or documentaries. Sometimes the QuickTime playback of streaming videos also displays this effect. Typically, a film needs to display 24 frames a second (called *frames per second* or *fps*) to look smooth. For video, 30fps is needed.

As in movies, an illusion of motion in the CG world is created by playing back a sequence of CG images. However, it is not enough to just make the models move—their motion should look realistic. We shall present the basic principles of animation that help achieve realism in motion. We begin our animation journey by exploring the realm of traditional animation. We will then see how these principles are carried over and applied to a 3D CG world as we build our own animated movie.

In this chapter you will learn

- Traditional (2D, hand-drawn) animation
- 3D animation
- Applying the principles of animation to achieve realistic motion

9.1 Traditional Animation

In traditional 2D animation, most of the work is done by hand. The story to be animated is written out on a storyboard. Initially only the *key frames* are identified. Key frames are frames selected by the master animators for their importance in the animated sequence, either because of the composition of the frames, an expression depicted by the characters, or an extreme movement in the sequence.

Let us reconsider the smiley face sequence we saw in Fig.9.1. The key frames of this sequence would be the frames with the smile and the final frown, as shown in Fig.9.2.

Key Frame #1

Key Frame #2

Fig.9.2: Key Frames in the smiley animation

Once the key frames are drawn, the next step is to draw the intermediate frames. The number of intermediate frames, or *in-betweens* as they are called, depends on the length of the animated sequence desired (keeping in mind the fps count that needs to be achieved). This step of drawing in-betweens is also called *tweening*.

If we were going to create a video of the face smiley animation, we would need to create 30 fps. For a short five minute animation, we would need (30 fps) x (60 seconds) x (5 minutes) = 9000 frames to be drawn! Now you know why Disney hires so many animators to work on a single movie.

A technique that helps tremendously in the process of creating animations, is called *cel animation*. When we create an animation using this method, each character is drawn on a separate piece of transparent paper. A background is also drawn on a separate piece of opaque paper. Then, when it is time to shoot the animation, the different characters are overlaid on top of the background. Each frame can reuse the same elements and only modify the ones that change from one frame to the next. This method also saves time since the artists do not have to draw in entire frames-just the parts that need to change, such as individual characters. Sometimes, even separate parts of a character's body are placed on separate pieces of transparency paper.

Traditional animation is still a very time-consuming and labor-intensive process. Additionally, once the frames are drawn, changing any parts of the story requires a complete reworking of the drawings.

How can we make animation less laborious and more flexible? The answer is found in the use of computers.

Computers are used by traditional animators to help generate in-betweens. The animators sketch the key frames, which are then scanned into the computer as line strokes. Many software programs are also available that enable drawing directly on the computer. The computer uses the line strokes of the key frames to calculate the in-betweens. Using sophisticated algorithms and equations, the in-betweening can be made to look very natural and smooth.

The use of computers is not, however, limited to this minor role. Since the advent of 3D Computer Graphics, computers are used in a big way to generate the entire animation sequence in a simulated 3D world. The sequence can be a stand-alone film, like *Toy Story* or *The Incredibles*, or can be composited into a real action film, as in *Lord of the Rings* or *Terminator-2*.

9.2 3D Computer Animation - Interpolations

The most popular technique used in 3D computer animation is the key frame in-betweening technique that was borrowed from traditional animation. In this technique, properties of a model (such as its position, color, etc.) are identified by a user (usually an animator) in two or more key frames. The computer then calculates these properties of the model in the in-between frames in order to create smooth motion on playback.

Computing the in-between frames is done by taking the information in the key frames and averaging it in some way. This calculation is called *interpolation*. Interpolation creates a sequence of in-between property values for the in-between frames. The type of interpolation used depends on the kind of final motion desired.

Interpolation can be used to calculate different properties of objects in the in-between frames-properties such as the position of objects in space, its size, orientation, color etc.

Let us learn some more about some basic interpolation methods and how we can use them to define an animation.

Linear Interpolation

The simplest type of interpolation is linear interpolation. In linear interpolation, the values of a given property is estimated between two known values on a straight-line basis (hence the term *linear*). In simpler terms, linear interpolation takes the sum of the property values in the two key frames and divides by the number of frames needed to provide as many equally spaced in-between frames as needed. For example, consider a model P, with a position along the y-axis of Y_{in} at the initial key frame K_{in}. At the final key frame, K_{fin}, the *y* position has moved to Y_{fin}. Now, say we want only one in-between frame f. This frame would be midway between the two key frames, and P would be located midway between the positions Y_{in} and Y_{fin} as shown in Fig.9.3.

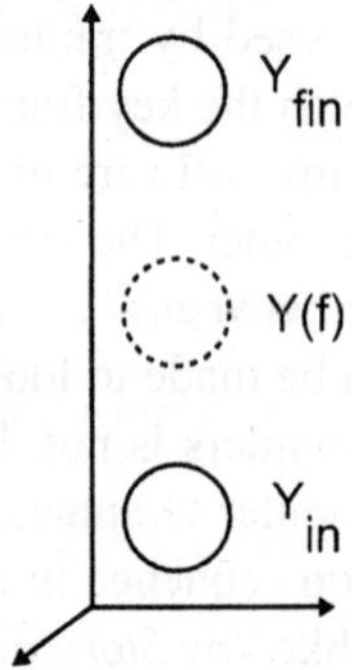

Fig.9.3: Linearly interpolating the Y position

Mathematically, the position Y(f) at Frame f, is given by

$$Y(f)=Y_{in} + (Y_{fin} - Y_{in}) / 2 = (Y_{fin} + Y_{in}) / 2$$

Now, suppose we want two in-between frames, f_1 and f_2. The model would then be defined such that at f_1, it occupies a position, Y(f1), which is one third the distance from Y_{in}; and at frame f_2, it occupies a position Y(f2), which is two thirds the distance from Y_{in}.

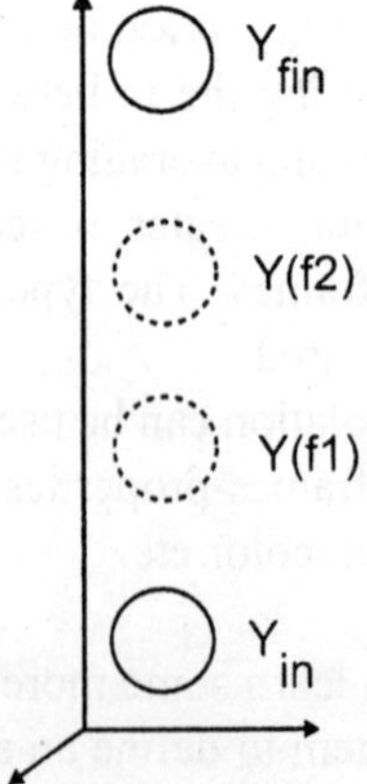

Fig.9.4: Two inbetween frames

Mathematically,

$$Y(f1) = Y_{in} + (Y_{fin} - Y_{in}) / 3$$

$$Y(f2) = Y_{in} + 2 * (Y_{fin} - Y_{in}) / 3$$

In total, we would have four frames: K_{in}, f_1, f_2, and K_{fin}—enough to make an animation that lasts for 1/6th of a second (assuming 24 frames per second).

Animations are usually represented in terms of timelines. The animation timeline defines the frames of the animation along an axis. The property values for the active object at these frames can be depicted along this axis either visually or just as values. For our example, we could define the following animation timelines:

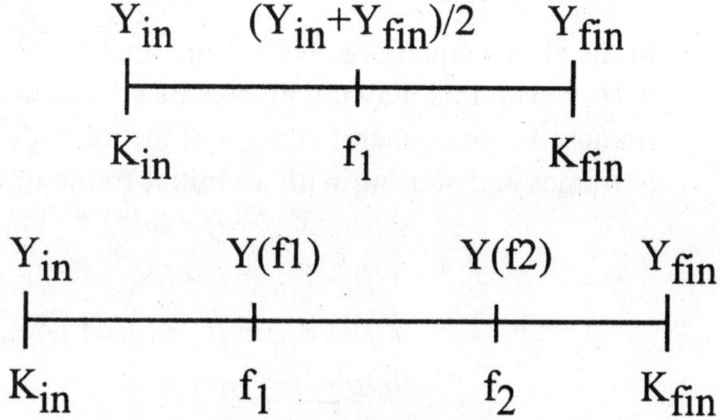

Fig.9.5: Animation timeline

If we were to extend our discussion to a total of *N* frames (including the initial and final key frames), we would have *N-2* in-between frames. The position Y at any frame *f*, can be can be calculated by the equation

$$Y(f) = Y_{in} + f^*(Y_{fin} - Y_{in}) / (N - 1)$$

$$0 \le f \le N - 1$$

At frame $f = 0$, we attain $Y(0) = Y_{in}$: the initial key frame.
At frame $f = N-1$, $Y(N-1) = Y_{fin}$: the final key frame.

The N-2 in-between frames can be calculated by varying f from 1 to (*N*-2). As we vary f, we linearly interpolate the position of P from Y_{in} to Y_{fin}.

Do you recognize this equation? Yes-it's the parametric equation of a line! The parameter we use is the frame number.

We can define similar equations for interpolating the position of P along the *x*- and *z*-axis:

$$X(f) = X_{in} + f^*(X_{fin} - X_{in}) / (N - 1)$$

$$Z(f) = Z_{in} + f^*(Z_{fin} - Z_{in}) / (N - 1)$$

$$0 \le f \le N - 1$$

This set of equation enables us to linearly interpolate the position of model P between two key frames. We can apply the same equations to interpolate rotation, scale, color, etc of the object. What we are doing is calculating the transformation at each in-between frame using a linear interpolation equation. The code to calculate the intermediate value is as shown below and can be found in the file *linearinterpolation.cpp*, under the installed directory for our sample code:

```
GLfloat InterpolatedValue(GLfloat KFinitvalue, GLfloat KFfinalvalue, int N,
int f){
    GLfloat ivalue;
    ivalue = KFinitvalue + f*(KFfinalValue-KFinitValue)/(N-1);
}
```

In the above equations, we assume that the first frame of the sequence is frame 0. However, this may not always be so, especially if we have more than two key frames. A more general version of the interpolation equation, assuming a total of N frames and starting with an initial frame of InitF is as follows:

$$X(f) = X_{in} + (f - Initf) * (X_{fin} - X_{in}) / (N - 1)$$
$$Y(f) = Y_{in} + (f - Initf) * (Y_{fin} - Y_{in}) / (N - 1)$$
$$Z(f) = Z_{in} + (f - Initf) * (Z_{fin} - Z_{in}) / (N - 1)$$
$$Initf \leq f \leq Initf + N - 1$$

This equation can be calculated by our function *InterpolatedValue*, if we pass in the parameter f as the value, *(currentframe-Initf)*

Example Time

Let us take the example of a bouncing ball to illustrate how linear interpolation works. We will initially move the ball only along the y-axis to grasp the concepts. Then we will expand the motion along the x-axis as well.

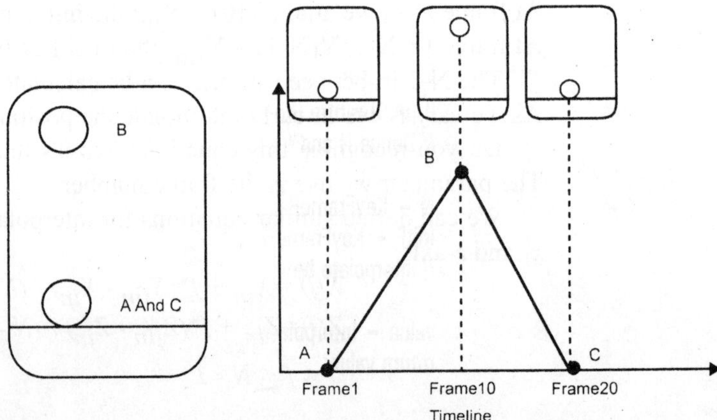

Fig.9.6 : Key frames and graph of a bouncing ball

As shown in Fig.9.6, a ball is initially at position A=(0,Y_a,0). It goes up to position B=(0, Y_b,0) and the falls back to Position C=(0, Y_a,0). The motion of the ball can be drawn as a graph of its position in Y against the animation timeline, also shown in Fig.9.6.

The position of the ball at A, B and C define our key frames. Let us say, we want our animation to be 20 frames long. Further, let the three key frames be at frames 1,10, and 20 along the animation timeline. To define the linear interpolation between positions A and B, we go through 10 frames from 1 to 10. The interpolated values of Y for any frame f is defined as

$$Y(f) = Y_a+(f-1)*(Y_b-Y_a)/(10-1)$$
$$1 \le f \le 10$$

In this same sequence, positions for frames between 10 and 20 need to be linearly interpolated between positions B and C. These positions can be expressed as

$$Y(f) = Y_b+(f-10)*(Y_c-Y_b)/9$$
$$10 < f \le 20$$

Using the two equations, we are able to define the position of the ball at any intermediate frame in the animation.To generalize the code to calculate these equations, assume we stored the positions of the key frames in an array called KeyFramePositions[]. In our example, KeyFramePositions = {0,10,20}. We store the property values in an array called KeyFrameValues[], which in this case = {Y_a, Y_b, Y_c }. Then we can define a function that can calculate the intermediate value at any given frame as

```
GLfloat EvaluateLinearAt(GLfloat *KeyFrameValues, GLint *KeyFramePositions, GLint
NumOfKeyFrames, GLint f)
{
        int i;
        GLint N, InitF;
        GLfloat value;

        // Find the two keyframes which this frame interpolates
        for (i-0;i<NumOfKeyFrames;i++){
                if (f < KeyFramePositions[i])
                        break;
        }
        i--;
        if (i<0) return (KeyFrameValues[0]); // should not happen
        // Define N and InitF

        N - KeyFramePositions[i+1]-KeyFramePositions[i];
        InitF - KeyFramePositions[i];
        // interpolate between KeyFramePositions[i] and KeyFramePositions[i+1]

        value - InterpolatedValue(KeyFrameValues[i], KeyFrameValues[i+1], N, (f-InitF));
        return value;

}
```

The code can also be found in the file provided: *linearinterpolation.cpp.* In *Example9_1*, we define a ball bouncing up and down. Its position at KeyFrame A is $Y_a = 0$; at B, $Y_b = 25$; and at C, $Y_c = 0$ again. We define

```
GLFloat *KeyFramePositions - {1,10,20}.
Glfloat *KeyFrameValues -  {0,10,0}
int NumberofKeyFrames - 3;
int MAXFRAMES - 20;
```

If we let the glut sphere represent our ball, then the code for this animation looks as follows:

```
ypos - EvaluateLinearAt(KeyFrameValues, KeyFramePositions,
NumberofKeyFrames, currentFrame);
 glLoadIdentity ();
 gluLookAt(0.,0.,13, 0,0,-100,0.,1.,0.);
 glTranslatef(0.,ypos,0.);
 glutWireSphere(.5,10,10);
```

where currentFrame is updated every loop by a timer callback. The entire code example code be found in *Example9_1/Example9_1.cpp.*

Let us make the ball bounce some more. Due to the nature of gravity, each time the ball bounces back up, it bounces to a point lower than the previous bounce. To accomplish this, we extend our animation to 70 frames, and add in a six more key frames at D, E, F, G, H, and I as shown in Fig.9.7. The only changes

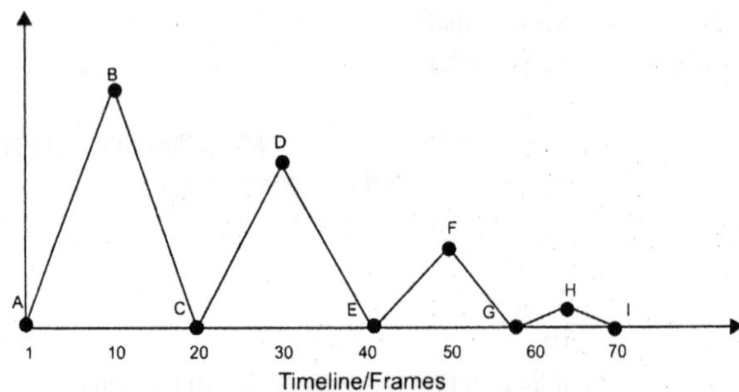

Fig.9.7: Key Frames for Bouncing Ball

you need to make to the *Example9_1* is to define the key-frame arrays as:

```
GLfloat KeyFrameValues[9] - {0,10,20,30,40,50,60,70};
GLint KeyFramePositions[9] - {0,10,0,7,0,4,0,1,0};
int NumberofKeyFrames - 9;
int MAXFRAMES - 71;
```

Voila! You will see a ball bouncing.

Let us now add in some motion along the x-axis, so that the ball appears to have been thrown by someone, and is slowly bouncing to rest.

In order to accomplish this, we define a set of key frames for motion along the *x*-axis as

```
KeyFrameXPosValues[0] - 0;
KeyFrameXPosValues[1] - 3.5;
KeyFrameXPosValues[2] - 6;
KeyFrameXPosValues[3] - 8;
KeyFrameXPosValues[4] - 9.5;
KeyFrameXPosValues[5] - 11;
KeyFrameXPosValues[6] - 12;
KeyFrameXPosValues[7] - 12.5;
KeyFrameXPosValues[8] - 13;
```

We now interpolate the ball's *x* position as well as the *y* position by calling the interpolation function

```
ypos - EvaluateLinearAt(KeyFrameValues, KeyFramePositions, NumberofKeyFrames,
currentFrame);
xpos - EvaluateLinearAt(KeyFrameXPosValues, KeyFramePositions, NumberofKeyFrames,
currentFrame);
 glLoadIdentity ();
gluLookAt(0.,0.,13, 0,0,-100,0.,1.,0.);
glTranslatef(xpos,ypos,0.);
glutWireSphere(.5,10,10);
```

This example can be found in *Example9_2/Example9_2.cpp*. You will see the animated ball moving along the x- and y-axis as defined.

In real life, we do not expect the ball to actually follow a straight-line trajectory along its path. Also, at the top, where the ball reaches its highest point, we expect the ball to slow down before it changes direction. The motion of our ball is a bit jerky and abrupt.

The key drawback of linear interpolation is that it is based on constant speed between key frames, and the interpolation defines straight-line paths. This makes the animation looks jerky. Let us see how nonlinear interpolation can rescue us from this predicament.

Let us see how non-linear interolation can rescue us from this predicament.

Non Linear Interpolation

For most animation situations, linear interpolation is not an effective animation technique. More complex non-linear interpolation techniques are employed for simulating effects in nature.

Recall the nonlinear paramateric equations from Chapter 7.

$$x(t) = a_x t^3 + b_x t^2 + c_x t + d_x$$
$$y(t) = a_y t^3 + b_y t^2 + c_y t + d_y$$
$$z(t) = a_z t^3 + b_z t^2 + c_z t + d_z$$
$$(0 \leq t \leq 1)$$

As useful as these equations are in modeling, they are even more useful in animation! Instead of interpolating vertex points of a model, we can interpolate key frame positions of an animation using the very same equations. The parametric cubic equation set can be replaced to use the frame, f, as the interpolating parameter:

$$X(f) = a_x f^3 + b_x f^2 + c_x f + d_x$$
$$Y(f) = a_x f^3 + b_x f^2 + c_x f + d_x$$
$$Z(f) = a_x f^3 + b_x f^2 + c_x f + d_x$$

$$f = (f\text{-}Initf)/(N\text{-}1)$$
$$Initf \leq f \leq N \Rightarrow 0 \leq f \leq 1$$

The coefficients define the type of cubic spline being used. We covered bicubic splines in Chapter 7. Bicubic splines are great for modeling, but they can be cumbersome to use, since they only approximate the two interior points. In this chapter, we shall use coefficients that specify a derivative of a hermite cubic spline called the *Cardinal cubic spline.* The advantage of this spline is that the curve interpolates all the points specified. We need at least four key frames in order to define a cubic interpolation. In the file, *cubicinterpolation.cpp*, we define a function to evaluate a Hermite cubic interpolation

```
GLfloat tension - 0.3;
GLfloat CardinalValue(GLfloat KFValuem1, GLfloat KFValue1, GLfloat KFValue2, GLfloat KFValuep2, int
N, int f)
{

        GLfloat F - 1.0*f/(N-1);

        GLfloat tangent1 - (1.-tension)*(KFValue2-KFValuem1)/2.;
        GLfloat tangent2 - (1.-tension)*(KFValuep2-KFValue1)/2.;

        GLfloat hvalue - F*F*F*(2.*KFValue1-2.*KFValue2+tangent1 + tangent2) +
                                               F*F*
(-3.*KFValue1+3.*KFValue2 -2.*tangent1 - tangent2) +
                                               F*(tangent1)+
                                               KFValue1;
        return hvalue;
}
```

The *tension* variable controls the tangents at the control points. Smaller values of tension tend to produce tighter curves-the motion at the control point is more abrupt. Increasing the value of tension will relax the curve at the control points, leading to slower motion around the control points.

For the bouncing ball from *Example9_2*, if we applied hermite interpolation, the animation plot would appear as follows:

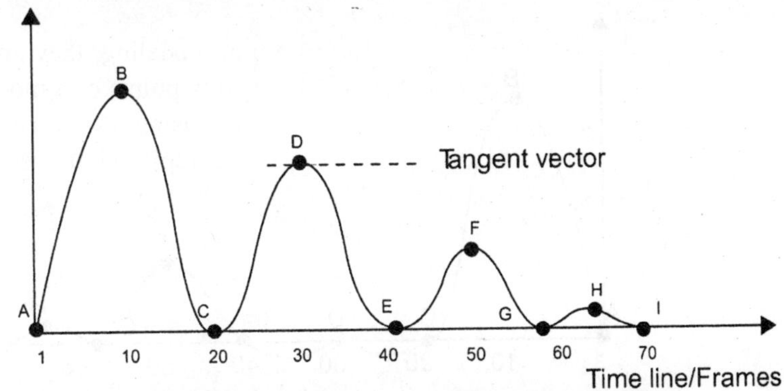

Fig.9.8: Hermite Interpolation Timeline

You can see how the ball's motion slows down at the control points.

In *cubicinterpolation.ccp*, we also define a function:

```
GLfloat EvaluateCubicAt(GLfloat *KeyFrameValues, GLint
```

This function figures out which four control points to send in to the CardinalValue function. It also appends a copy of the first and last points to our keyFrameValues array to interpolate the first and last points correctly. In *Example9_3*, we modify our ball animation to now use the cubic interpolation:

```
ypos - EvaluateCubicAt(KeyFrameValues, KeyFramePositions,
NumberofKeyFrames, currentFrame);
xpos - EvaluateCubicAt(KeyFrameXPosValues, KeyFramePositions, NumberofKeyFrames,
currentFrame);
glLoadIdentity ();
gluLookAt(0.,0.,13, 0,0,-100,0.,1.,0.);
glTranslatef(xpos,ypos,0.);
glutWireSphere(.5,10,10);
```

Watch the ball bounce. Does the animation look smoother? We achieved the desired slow-down of the ball at the top peaks. However, do we really want the same effect at the bottom? The ball should actually speed up at the bottom! Try changing the value of the tension parameter to see how the motion is affected at the endpoints. Most animation software employs equations that let the user interactively tweak the interpolation graph at every control point in order to achieve desired behavior.

There are cases when cubic interpolation does not work too well. Consider a situation where, in key frame D, we want the ball to stay on the ground as shown in Fig.9.9. The graph shows a wiggle between points C, D and E that makes the graph go below the 0 level we defined for the ball. If we had the ball animating

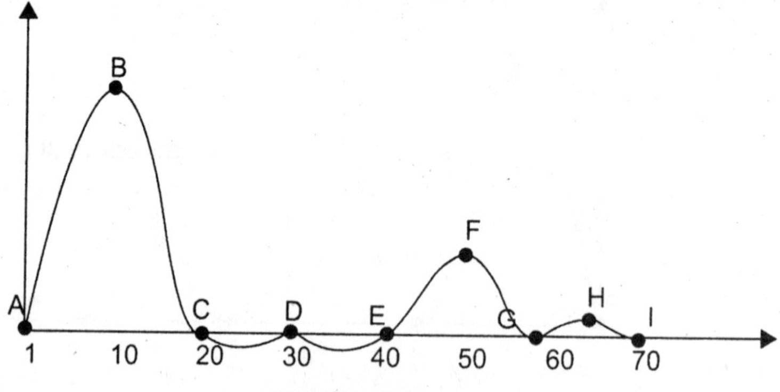

Fig.9.9: A wiggle below the ground

on top of a ground plane, then the ball would go underneath it! We would need to add in extra control points to avoid this behavior. Try this with your ball animation by changing the KeyFrameValues array and see what happens. Try it for the linear interpolation case as well. Unlike linear interpolation, cubic interpolation does not retain the values of adjacent key frames when these are equal.

It is hard to achieve accurate realism using only one interpolation technique. In most cases, a combination of different schemse is needed to get good results. Extensive ongoing research constantly delivers better ways to make realistic interpolations possible. In any production, the animator has a big role in understanding how interpolation works and how to modify the interpolation graphs to achieve desired affects.

Animating Snowy

Let us use our interpolation equations to create a slightly more complicated animation. Recollect our snowman model from Chapter 5. We shall use it to create an animation of a snowman bouncing on the ground.

In *Example9_4*, we reuse the snowman model from Chapter 5. We changed the snowman model code to comment out transforms that we did not need - for sake of speedier display. Again for speed considerations, we shall preview our snowman animation in wire frame mode.

In this example, we will only animate the root level component of the snowman - the null node. Recall that the null node is the parent of all the components of the snowman. Its pivot point is at the base of the snowman. Recall that this node (as well as each component of the snowman) had a transform array applied to it. These transforms were specified in an array of size 9 storing the (Tx,Ty,Tz,Sx,Sy,Sz,Rx,Ry,Rz) values:

```
GLfloat snomanXforms[9] = {0.,0.,0.,0.,0.,0.,1.,1.,1.};
GLfloat botXforms[9] = {0.,0.,0.,0.,0.,0.,1.,1.,1.};
```

In this example, instead of using a one-dimensional array of the nine xform values, we use a two dimensional array, to hold the 9 xform values for each *key frame* we define for the snowman (we assume a max of 10 key frames) This leads us to define the following structure to set the xform values for the snowman at its key frames:

```
GLfloat smKeyFrameValues[10][9]; // [frame][xform values]
```

The array to hold the key frame positions and number of key frames defined for the snowman are defined as follows:

```
GLint smKeyFramePositions[10];
//This array holds the actual positions of the key frames defined for the snowman
GLint smNumKeyFrames;
//The number of key frames that the snowman has
```

We define a utility function to actually initialize the key frame values for the snowman.

```
        // Set the key frame at frame=kf with the specified transforms
void SetKeyFrame(int kf, GLfloat keyframevalues[15][9], GLint
*keyframepositions, GLint *numKF,GLfloat Tx, GLfloat Ty, GLfloat Tz,
GLfloat Rx, GLfloat Ry, GLfloat Rz ,GLfloat Sx, GLfloat Sy, GLfloat Sz)
{
        keyframepositions[*numKF] = kf;
        keyframevalues[*numKF][0] = Tx;
        keyframevalues[*numKF][1] = Ty;
        keyframevalues[*numKF][2] = Tz;
        keyframevalues[*numKF][3] = Rx;
        keyframevalues[*numKF][4] = Ry;
        keyframevalues[*numKF][5] = Rz;
        keyframevalues[*numKF][6] = Sx;
        keyframevalues[*numKF][7] = Sy;
        keyframevalues[*numKF][8] = Sz;
// increment number of keyframes defined for model
        *numKF = *numKF+1;
}
```

To evaluate the position of the snowman at any intermediate frame, we define a function

```
void EvaluateXformsAt(GLfloat KeyFrameValues[10][9], GLint *KeyFramePositions, GLint NumOfKeyFrames, GLint f, GLfloat *xforms)
{
    // interpolate between KeyFramePositions[i] and KeyFramePositions[i+1]
    N - KeyFramePositions[i+1]-KeyFramePositions[i]+1;
    InitF - KeyFramePositions[i];

    xforms[0] - LinearValue(KeyFrameValues[i][0], KeyFrameValues[i+1][0], N, (f-InitF));
    xforms[1] - LinearValue(KeyFrameValues[i][1], KeyFrameValues[i+1][1], N, (f-InitF));
    xforms[2] - LinearValue(KeyFrameValues[i][2], KeyFrameValues[i+1][2], N, (f-InitF));
    xforms[3] - LinearValue(KeyFrameValues[i][3], KeyFrameValues[i+1][3], N, (f-InitF));
    xforms[4] - LinearValue(KeyFrameValues[i][4], KeyFrameValues[i+1][4], N, (f-InitF));
    xforms[5] - LinearValue(KeyFrameValues[i][5], KeyFrameValues[i+1][5], N, (f-InitF));
    xforms[6] - LinearValue(KeyFrameValues[i][6], KeyFrameValues[i+1][6], N, (f-InitF));
    xforms[7] - LinearValue(KeyFrameValues[i][7], KeyFrameValues[i+1][7], N, (f-InitF));
    xforms[8] - LinearValue(KeyFrameValues[i][8], KeyFrameValues[i+1][8], N, (f-InitF));
```

This function figures out the appropriate key frames to interpolate between and then calls the linear interpolation function to evaluate the transform for the current frame. This transform is then stored in the parameter array: *xforms*. Let us bounce Snowy up and down on a snowy ground. We define our world with it's extents being about -2 to 15 units along the x- and 0 to 12 unit along the y-axis. We define a snowy ground plane as a solid cube that is grayish in color.

```
void draw_Ground(){
    glPolygonMode(GL_FRONT, GL_FILL);
    glPolygonMode(GL_BACK, GL_FILL);
    glColor3f(0.8,0.8,0.8);
    glPushMatrix();
    glTranslatef(0.,-5.5,0.);
    glScalef(25.,10.,10.);
    glutSolidCube(1.);
    glPopMatrix();

    glPolygonMode(GL_FRONT, GL_LINE);
    glPolygonMode(GL_BACK, GL_LINE);
}
```

The ground is based at y=0, but does not extend all the way to the end of the world coordinates. We will define 45 frames in this animation sequence. We want to open the sequence with Snowy jumping into the scene. To do this, we define a key frame for Snowy at frame 0, with his Tx=-5 and Ty =6.

```
SetKeyFrame(0,smKeyFrameValues, smKeyFramePositions, &smNumOfKeyFrames, -5.,6.,0., 0.,0.,0.,
1.,1.,1.);
```

At frame 15, we want Snowy to hit the ground.

```
SetKeyFrame(15,smKeyFrameValues, smKeyFramePositions, &smNumOfKeyFrames,
4.,0.,0.,1.,1.,1.,0.,0.,0.);
```

At frame 30, Snowy rises back up again, and at frame 45 he lands back down-but oh no! This time he has not landed back on the ground!

```
SetKeyFrame(30,smKeyFrameValues, smKeyFramePositions, &smNumOfKeyFrames,
10.,6.,0.,1.,1.,1.,0.,0.,0.);
SetKeyFrame(45,smKeyFrameValues, smKeyFramePositions, &smNumOfKeyFrames,
14.,0.,0.,1.,1.,1.,0.,0.,0.);
```

A timer function periodically updates the current frame number and calls the Display function. The Display function merely calls the EvaluateXformsAt to evaluate the transforms of the snowman for the current frame and calls the snowman drawing code with these transforms.

```
void Display(void)
{
    GLfloat ypos, xpos;
    glutSwapBuffers();

    glClear(GL_COLOR_BUFFER_BIT | GL_DEPTH_BUFFER_BIT);

    EvaluateXformsAt(smKeyFrameValues, smKeyFramePositions, smNumOfKeyFrames,
currentFrame, snomanXforms);
  draw_SnowMan(snomanXforms, botXforms, tumXforms, headXforms, lEyeXforms, rEyeXforms,
noseXforms, lHandXforms, rHandXforms);

    draw_Ground();

    glFlush();
}
```

Remember the use of the Push and Pop matrix-this ensures that when we pop back from the snowman drawing routine, we are back in the transformation state that we started with. So we do not need to adjust the position of the ground.

We have the basic motion of Snowy down. If you run the program, you will notice that Snowy doesn't look very natural just facing us while he is bouncing. To make him look in the direction he is going add in a rotation about y of about 40 degrees. View your animation. We want to add more to this animation. We would like Snowy to tilt back as he lands and reorient himself for takeoff. To do this, we redefine the key frames as follows. At frame 0, we just tilt Snowy back by rotating him by 20 degree around the Z axis.

```
SetKeyFrame(0,smKeyFrameValues, smKeyFramePositions, &smNumOfKeyFrames,
5.,6.,0., 0.,40.,20.,1.,1.,1.);
```

At frame 10, he lands on the ground:

```
SetKeyFrame(10,smKeyFrameValues, smKeyFramePositions, &smNumOfKeyFrames,
4.,0.,0., 0.,40.,20. ,1.,1.,1.);
```

At frame 13, he straightens himself up:

```
SetKeyFrame(13,smKeyFrameValues, smKeyFramePositions, &smNumOfKeyFrames,
4.,0.,0., 0.,40.,0,1.,1.,1.);
```

At frame 16, he reorients himself for take off:

```
SetKeyFrame(16,smKeyFrameValues, smKeyFramePositions, &smNumOfKeyFrames,
4.,0.,0., 0.,40.,-20,1.,1.,1.);
```

At frames 26 and 30 Snowy reaches the highest point of his bounce and re-aligns his body:

```
SetKeyFrame(26,smKeyFrameValues, smKeyFramePositions, &smNumOfKeyFrames,
8.,6.,0.,1.,1.,1.,0.,40.,-20.);
SetKeyFrame(30,smKeyFrameValues, smKeyFramePositions, &smNumOfKeyFrames,
10.,6.,0.,1.,1.,1.,0.,40.,20.);
```

In frame 40, Snowy lands (but not on the ground); we clamp his motion on frame 45 so he won't move any farther for the rest of the animation:

```
SetKeyFrame(40,smKeyFrameValues, smKeyFramePositions, &smNumOfKeyFrames,
15.,0.,0.,1.,1.,1.,0.,40.,20.);
SetKeyFrame(45,smKeyFrameValues, smKeyFramePositions, &smNumOfKeyFrames,
15.,0.,0.,1.,1.,1.,0.,40.,0.);
```

View your animation. You can add in surface properties to the snowman and light up your scene to see a solid bouncing snowman.

The code can be found under *Example9_4* in files, *Example9_4.cpp*, *Smowman.cpp* and *Snowman.h*. The animation looks good; but it can still do with some zing!

9.3 The Principles of Animation

Three-dimensional computer animation has made great advances in the last few years. There are numerous animation packages (we shall look into Maya in the next chapter) that can help even a novice develop quality animations. However, it is not uncommon to see animations that lack zing-they seem drab, look unreal and reek of CGism. Here is where the principles of traditional 2D animation come to the rescue.

The concepts used in hand-drawn animations, to make the sequence look believable, can be used in the 3D realm to add as much zest and depth into the animation as we see in its 2D counterpart.

Walt Disney himself put forward many of these principles of animation which is now used as a bible by animators worldwide. Refer to LASS87 for a more detailed analysis of these principles.

In this section, we shall explore some of these principles and see how we can apply them to Snowy, whom we animated earlier. We will progressively add more key frames into our sequence from *Example9_4,* to enhance the appeal of the bouncing Snowman. The (entire) modified code can be found under *Example9_5/Example9_5.cpp.*

Squash and Stretch

In real life, only the most rigid objects retain their original shape during motion. Most objects show considerable movement in their shapes during an action. If you notice how a ball bounces on the ground, it seems to squash on impact and stretch back up when rising. The amount of squash depends on how rigid the ball is-a soft rubber ball squashes a lot more than a hard baseball.

One trick used in animation to depict the changes in shape that occur during motion is called *squash and stretch*. The *squashed* position depicts a model form flattened out by an external pressure or constricted by its own power. The *stretched* position shows the same form in a very extended or elongated condition. Squash and stretch effects also help to create illusions of speed and rigidity.

Human facial animation also uses the concepts of squash and stretch extensively to show the flexibility of the skin and muscle and also to show the relationship between parts of the face. During the squash and stretch, the object should always retain its volume. If an object squashes down, then its sides should stretch out to maintain the volume.

The standard example of this principle is the bouncing ball. Fig.9.9 shows some of the frames in a bouncing ball example. The ball is squashed once it has made impact with the ground. This gives the sense of the material of the ball. The softer the ball, the greater the squash. The ball is elongated before and after it hits

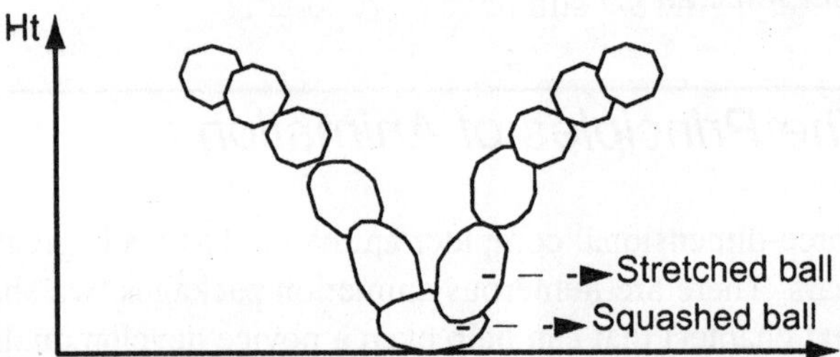

Fig.9.10: Squash and Strecth applied to a bouncing ball

the ground to enhance the feeling of speed. Hence, squash and stretch effects help to create illusions of speed and rigidity. Its one of the most important principle used in animation to simulate realism. Human facial animation also use the concepts of squash and stretch extensively to show the flexibility of the skin and muscle and also to show the relationship between parts of the face. We will use the principles of squash and stretch to add some life to our Snowy animation.

We will be using the scaling transforms to add squash and stretch to Snowy. At frame 10, Snowy is just about to land on the ground. Let us stretch him up at this frame to enhance the feeling of him speeding up in motion before impact with the ground. To do this, set $S_y = 1.2$. Since we wish to retain volume, we will squash him down in x- by setting $S_x = 0.8$:

```
SetKeyFrame(10,smKeyFrameValues,     smKeyFramePositions,
&smNumOfKeyFrames, 4.,0.,0., 0.,40.,20.,0.8,1.2,1.);
```

At frame 13, Snowy has made maximum impact with the ground. He should be squashed to depict this force. Set Sx = 1.2 and Sy = 0.8

```
SetKeyFrame(13,smKeyFrameValues,     smKeyFramePositions,
&smNumOfKeyFrames, 4.,0.,0., 0.,40.,0.,1.2,0.8,1.);
```

Finally, at frame 16, Snowy is ready to launch back into the air. Set his Sy = 1.2 and Sx = 0.8 to give the illusion of him stretching back up:

```
SetKeyFrame(16,smKeyFrameValues, smKeyFramePositions, &smNumOfKeyFrames,
4.,0.,0., 0.,40.,-20.,0.8,1.2,1.);
```

At frame 40, we will make Snowy stretch as if he is about to land on the ground (even though he really isn't!)

```
SetKeyFrame(40,smKeyFrameValues,          smKeyFramePositions,
&smNumOfKeyFrames, 15.,0.,0., 0.,40.,20,0.8,1.2,1.);
```

View your animation. Does the animation start looking more realistic already? And we still have more to go!

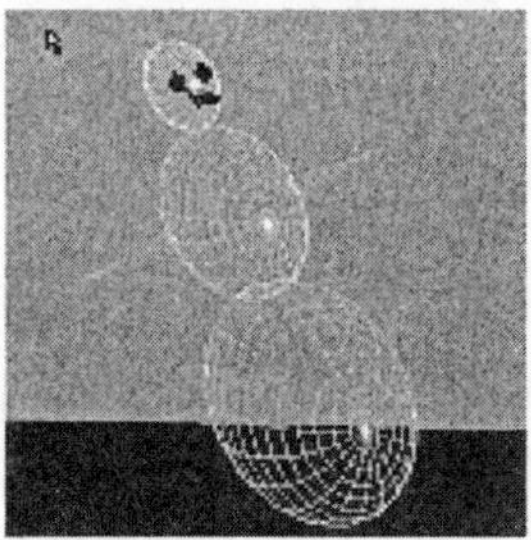
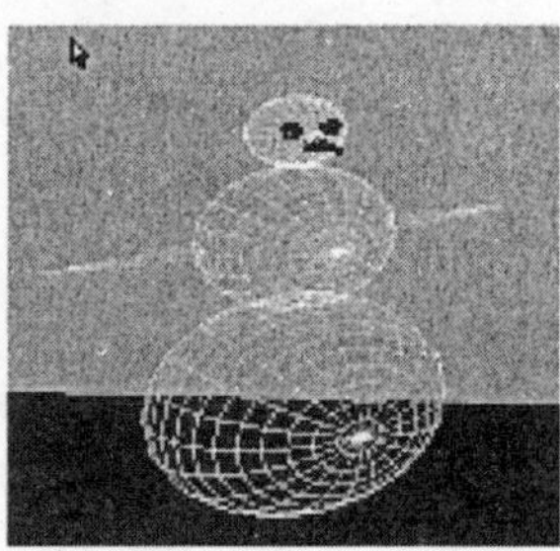

Fig.9.11: snowy: Squashed and then ready ot bounce up again!

Staging

Another important concept in animation is *staging*. Staging is the presentation of an idea so that it is completely and unmistakably clear. A personality is staged so that it is recognizable; an expression or movement is brought into focus so that it can be seen. It is important to focus the audience on only one idea at a time. If a lot of action is happening on the screen at once, the eye does not know where to look, and the main idea will be upstaged. The camera placement is very important to make sure that the viewer's eye is led exactly to where it needs to be at the right moment. Usually, the center of the screen is where the viewer's eyes are focused. Lighting techniques also help in staging.

Let us apply staging to our Snowy animation. We shall make Snowy complete his second bounce only to find himself off the plateau and standing on thin air! To stage this part of the animation, we want to move the camera to zoom in on Snowy's current location. Such a swing in the camera is called a *cut*.

A cut splits a sequence of animation into distinctive shots. The transition from shot to shot can be a direct cut, which we shall use in this example. It could also be a smooth transition, which is something we will look into in a later chapter. Usually, the last few frames from the first shot are repeated in the second shot to maintain continuity between the shots. Let us implement these concepts to refine our Snowy animation. First, we increase the total number of frames in our animation:

```
int MAXFRAMES = 61;
```

The cut occurs at frame 41, when Snowy just lands on thin air. We can implement this cut by simply moving our camera position in the *Display* function as follows:

```
if (currentFrame == 41) { // cut
        glLoadIdentity();
        gluLookAt(16,5.,12, 0,0,-100,0.,1.,0.);
}
else if (currentFrame == 0) { // orginal location of camera
        glLoadIdentity();
        gluLookAt(8,7.,17, 0,0,-100,0.,1.,0.);
}
```

How would you implement the camera repositioning in the case of a smooth transition between the shots? Snowy himself needs more key frames defined. We had to increase the size of all the key frame arrays to accommodate for these new key frames.

```
GLfloat smKeyFrameValues[15][9];
GLint smKeyFramePositions[15];
```

We also modified the parameter to the *SetKeyFrame* function appropriately. Frames 41 onwards forms cut2 of the sequence. From frames 41 to 50, we repeat the last bits of the animation from shot1:

```
SetKeyFrame(41,smKeyFrameValues, smKeyFramePositions, &smNumOfKeyFrames,
12.5,3.5,0., 0.,40.,20,0.9,1.1,1.);
SetKeyFrame(47,smKeyFrameValues, smKeyFramePositions, &smNumOfKeyFrames,
15.,0.,0., 0.,40.,20,0.8,1.2,1.);
SetKeyFrame(50,smKeyFrameValues, smKeyFramePositions, &smNumOfKeyFrames,
15.,0.,0., 0.,40.,0.,1.,1.,1.);
```

We clamp the animation from frames 50 to 60 for now:

```
SetKeyFrame(60,smKeyFrameValues, smKeyFramePositions,
&smNumOfKeyFrames, 15.,0.,0., 0.,40.,0.,1.,1.,1.);
```

You can go ahead and view your animation or wait until we add in more features. (Yes, we agree: the cut looks jerky! We will show you how cuts are made smooth in the next chapter)

Anticipation

Anticipation involves preparing the objects for the event that will occur in the later frames of the animation sequence. If there is no anticipation of a scene, the scene may look rather abrupt and unnatural. The anticipation principle makes the viewer aware of what to expect in the coming scene and also makes the viewer curious about the coming action. Anticipation is particularly necessary if the event in a scene is going to occur very quickly and it is crucial that the viewer grasp what is happening in the scene.

Continuing our Snowy animation, we already have the stage set up for the audience: they have seen Snowy land on thin air. To create more anticipation, we shall make Snowy look down to give an impression that he is trying to fig. out what he is landing on. This would draw a viewer's attention to the fact that good old Snowy is going to have a free fall from this frame onwards.

The key frames for the head from frames 1 to 45 need to be clamped to their identity values:

```
SetKeyFrame(0,sheadKeyFrameValues, sheadKeyFramePositions,
&sheadNumOfKeyFrames, 0.,0.,0., 0.,0.,0.,1.,1.,1.);
SetKeyFrame(47,sheadKeyFrameValues, sheadKeyFramePositions,
&sheadNumOfKeyFrames, 0.,0.,0., 0.,0.,0.,1.,1.,1.);
```

From frames 47 to frame 55, Snowy looks down to see why he hasn't squashed:

```
SetKeyFrame(55,sheadKeyFrameValues, sheadKeyFramePositions,
&sheadNumOfKeyFrames, 0.,0.,0., 40.,-20.,0.,1.,1.,1.);
SetKeyFrame(60,sheadKeyFrameValues, sheadKeyFramePositions,
&sheadNumOfKeyFrames, 0.,0.,0., 40.,-20.,0.,1.,1.,1.);
```

Let us wait to finish our animation before finally viewing it.

Timing

Timing refers to the speed of an action. It gives meaning to the movement in an animation. Proper timing is crucial to make ideas readable. Timing can define the weight of an object: a heaver object is slower to pick up and lose speed than a lighter one. Timing can also convey the emotional state of a character: a fast move can convey the sense of shock, fear, apprehension, and nervousness, while a slow move can convey lethargy or excess weight.

In the Snowy animation, we let Snowy look down lazily, using 8 frames to look down (from frame 47 to frame 55). However, when he realizes his coming plight (free fall) we want him to look up in shock. We shall make him look up in only two frames to convey this sense. We also define his eyes to enlarge to exaggerate the sense of shock:

```
SetKeyFrame(55,sheadKeyFrameValues, sheadKeyFramePositions,
&sheadNumOfKeyFrames, 0.,0.,0.,40.,-20.,0.,1.,1.,1.);
SetKeyFrame(57,sheadKeyFrameValues, sheadKeyFramePositions,
&sheadNumOfKeyFrames, 0.,0.,0., 0.,-20.,0.,1.,1.,1.);
```

```
SetKeyFrame(57,seyeKeyFrameValues, seyeKeyFramePositions,
&seyeNumOfKeyFrames, 0.,0.,0., 0.,0.,0.,1.4,1.4,1.2,);
```

Almost there-one last trick and we will be able to view a very realistic animation.

Secondary Action

A secondary action is an action that results directly from another action. Secondary actions are important in heightening interest and adding a realistic complexity to the animation.

For Snowy, a secondary action would be his hands moving up and down as he bounces up and down. This motion of his hands would help him maintain his balance and also add some realism to his bouncing. His hands would move down when he is landing and raise up as he bounces up. When Snowy finally falls, his hands swing upwards to convey the sense of free fall that he is in. The key frames to rotate the hand upwards as Snowy falls down are shown below. We leave the rest of the secondary motion as an exercise for you.

```
/// keyframes for the left hand
    SetKeyFrame(62,slHandKeyFrameValues,
slHandKeyFramePositions,        &slHandNumOfKeyFrames, 0.,0.,0.,1.,1.,1.,0.,0.,0.);
    SetKeyFrame(65,slHandKeyFrameValues,
slHandKeyFramePositions,        &slHandNumOfKeyFrames, 0.,0.,0.,1.,1.,1.,0.,0.,-90.);
    SetKeyFrame(70,slHandKeyFrameValues,
slHandKeyFramePositions,        &slHandNumOfKeyFrames, 0.,0.,0.,1.,1.,1.,0.,0.,-90.);
```

Make the animation and play it. The code can be founder under *Example9_5*, in files *Example9_5.cpp, Snowman.cpp* and *Snowman.h.*

Watch the animation and take special note of the tricks you have used in this section to make the animation more fun and realistic. Particularly notice the change in camera viewpoint, the roll of Snowy's head, his eyes widening, and his hand movements, too. Interested readers should also try changing the interpolation technique for Snowy to be non-linear. What do you see? Is the motion better or worse than its linear counterpart? Add shading and lighting to your animation as well for a cool animation.

There are several other animation principles that can be applied to achieve realistic motion. Exaggeration or deliberate amplification of an idea/action is often employed to focus the viewer's attention. Another common technique used for achieving a subtle touch of timing and movement is called *slow in and slow out*. In this the object slows down before reaching a key frame. We saw how this makes for a natural look of the bouncing ball when it reaches its highest point. The use of all these animation principles, along with good presentation of a theme, can help produce eye-catching and entertaining animation.

9.4 Advanced Animation Techniques

Interpolation is just one of the techniques used to generate realistic animations. Depending on the effects desired, different kinds of animations can be combined to achieve desired effects. We discuss some of the more advanced animation techniques briefly in this section.

Dynamics

Dynamics is the branch of physics that describes how an object's physical properties (such as its weight, size, etc) are affected by forces in the world that it lives in (like gravity, air resistance, etc). To calculate the animated path, an object traversing the virtual 3D world is imbued with forces that model after the physics of the real world. These forces act on the models causing some action to occur. For example, blades of grass sway in the wind depending on the speed of the wind, and to a lesser extent, depending on the size of the blade.

In the 3D world, a wind force can be defined which when applied to models of grass, affects their transformation matrix in a manner defined by the laws of physics.. This kind of simulation was used in *A Bug's Life*, where an infinite number of grass blades were made to constantly sway in the background by defining wind forces on them. The foreground animation of Flick and his troop was still defined using Key Frame animation.

Procedural motion

In this technique, a set of rules is defined to govern the animation of the scene. Particle systems are a very good example of such a technique. Fuzzy objects like fire and dust or even water can be represented by a set of particles. These particles are randomly generated at a defined location in space, follows a certain motion trajectory, and then finally die at different times. The rules define the position, motion, color and size of the individual particle. A collection of hundreds of

Fig.9.12: Particles representing a water fall

particles can lead to realistic effects of fire, explosions, waterfalls, etc.

Motion Capture

Real-time motion capture is an advanced technique that allows animators to capture real-live motion with the aid of a machine. An actor is strapped with electronic nodes that transmit signals to the computer when the actor moves. The computer records this motion. The captured motion data can then be applied to the transformation matrix of the computer-generated characters to achieve a real -live effect.

Kinematics

Kinematics is the study of motion independent of the underlying forces that produce the motion. It is an important technique used to define articulated fig. animations.

Usually, a defined model is applied that has a skeleton structure consisting of a hierarchy of rigid links (or bones) connected at joints. Each link corresponds to certain patches of the model surface, also called the model *skin*. A joint can be rotated, causing the corresponding surface and the attached children bones to rotate about it. There are usually some constraints on how much each joint can be rotated.

This motion corresponds closely with the human anatomy, and hence is used very often for human like animations.

Consider the model of a leg, with a defined skeleton consting of three links L1, L2, and L3 at joints J1, J2 and J3 as shown in Fig.9.13. J1 is the head of the hierarchy. A rotation about J1 causes L1-J2-L2-J3-L3 (and the associated skin) to rotate as shown in Fig.9.14. You can also rotate about joints J2 or J3 to achieve a desired position.

Can you see the relation between this organization of joints and links and the human (or any other animals) leg? Can you visualize how this technique could

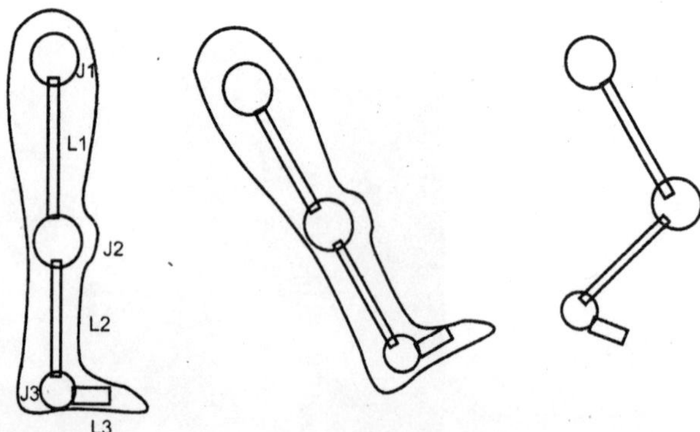

Fig.9.13: Rotation of the skelecton joints and links

be applied to our android? If you recall the example of the marching android, a similar principle was used to rotate his legs. He just didn't have any skin.

In forward kinematics, the motion of all the joints are specified explicitly by the animators. As the joints are rotated, the motions of the links in the chain are determined indirectly.

In inverse kinematics, the animator only specifies the final position of the links at the end of the hierarchy. Math equations are then used to détermine the orientation of all the joints in the path that lead to this link, with the rotation constraints applied against each joint.

Summary

In this chapter, we have looked at a few of the common animation techniques used to bring 3D models to life. We have seen the classic example of a bouncing ball and how different interpolation techniques yield different animations. We looked at some advanced animation techniques that can make the life of an animator simpler.

3D animation can be very drab if we do not apply the principles of traditional animation. We have studied these principles and used them to animate Snowy, our friendly snowman. In the next chapter, we will see how to animate the camera to produce a stunning game. In Chapter 11, we will create a small movie using Maya.

Chapter 10

Viewpoint Animation

In the last chapter, we learned some tricks to make our animations look more realistic and appealing. In all the previous examples, the position of the camera or our viewpoint was fixed. We only did a one-time change to the camera position: a cut in the animation sequence. Camera movement is not limited to changing across shots only. A technique often employed in animation is to move or transform the camera itself through a 3D world. This technique is called *viewpoint* or *camera animation*.

Why would one want to animate the camera? Well, camera movement adds valuable depth to the scene. Moving the camera through space causes objects to move in different ways depending on their distance from the camera, cluing us in to how they are spatially arranged in our 3D world space.

In games, the camera represents the player's eye: he can change his or her viewpoint by moving the joystick or the keyboard. A lot of planning is required to move the camera in a game. The speed, field of vision, and orientation of the camera are critical because what the camera sees is what an observer would perceive. The faster the camera moves the faster the observer feels he or she is moving. The sense of speed due to change in position and orientation obtained by camera motion can produce eye-catching and exciting effects. These techniques are commonly used in games and motion ride simulators, where you feel like you are moving through a real-life scene. Other spectacular examples of viewpoint animation can be found in animations of the solar system, the motion of atomic particles in a virtual model, etc.

What you will learn in this chapter:

- How to move the camera through space
- How to capture keyboard input to change camera positions
- How to implement your own 3D game complete with camera motion and fog

Let us begin this chapter by seeing how to move the camera in our snowy animation.

10.1 Animating the Camera in the Snowy Animation

The camera/viewpoint defines how the 3D scene is being viewed. The position of the camera and its orientation defines the final image rendered onto the screen. The position and orientation of the camera can be captured at certain key frames. To move the camera, we interpolate these camera parameters for the intermediate frames in exactly the same way we used interpolation for object transforms. For each in-between frame, the camera transformation matrix is computed and then used (in conjunction with the object transformation matrix) to compute the final scene orientation.

Let us see how to animate the camera with the snowman example from the previous chapter.

You can use two techniques to set the camera position: one is by using the call to

```
glulookAt
```

and the other is by making calls to the gl transformation functions (glTranslate, glScale etc) but in the opposite direction of the intended motion. Recall from Chapter 5, that moving the camera is the same transformation as moving the objects in the opposite direction. In OpenGL, the camera viewing matrix and the model transformation matrix are combined into one for this reason.

We shall see how to animate the camera using both techniques. Which technique you use is depends on how easy it is for you to think of your 3D world and how to move within it. Camera transforms should be the last transformations to be applied to the matrix stack, since they are applied to all the objects. This means that the camera transforms should be issued first. Recall how we cut between the two shots in the snowman animation in *Example9_5*:

```
if (currentFrame = = 41) { // cut
        glLoadIdentity();
        gluLookAt(16,5.,12, 0,0,-100,0.,1.,0.);
}
        else if (currentFrame = = 0) {*/
        glLoadIdentity();
        gluLookAt(8,7.,15, 0,0,-100,0.,1.,0.);
}
```

In *Example10_1*, instead of cutting between the two shots, we shall smoothly pan the camera across, starting at a viewpoint position of (8,7,15) at frame 30, and ending at the final position of (16,5,12) at frame 50. We do not animate the orientation (the look at point of (0,0,-100) and the normal vector of (0,1,0)) in this example. For this example, we define a set of key frames to hold the camera position from frames 0 to 30 at the initial camera position:

```
SetKeyFrame(0,camKeyFrameValues, camKeyFramePositions, &camNumOfKeyFrames,
8.,7.,17.,1.,1.,1.,0.,0.,0.);
SetKeyFrame(30,camKeyFrameValues, camKeyFramePositions, &camNumOfKeyFrames,
8.,7.,17.,1.,1.,1.,0.,0.,90.);
```

For the next 20 frames, the camera moves smoothly to its new position:

```
SetKeyFrame(50,camKeyFrameValues, camKeyFramePositions, &camNumOfKeyFrames,
16.,5.,12.,1.,1.,1.,0.,0.,90.);
```

To prevent any drift between frames 50 and 70, we define an extra key frame at frame 70:

```
SetKeyFrame(70,camKeyFrameValues, camKeyFramePositions, &camNumOfKeyFrames,
16.,5.,12.,1.,1.,1.,0.,0.,90.);
```

Finally, in the Display code, we interpolate the camera position and set the value accordingly. Note that we only interpolate the position of the camera.

```
EvaluateXformsAt(camKeyFrameValues, camKeyFramePositions, camNumOfKeyFrames,
currentFrame, camXforms);
glLoadIdentity();
gluLookAt(camXforms[0],camXforms[1],camXforms[2], 0,0,-100,0.,1.,0.);
```

Voila! Watch your snowman animation, with a smooth camera pan just before Snowy falls from the cliff. Does it help the animation in any way? Already you get a sense of perspective between Snowy and the plateau on which he is jumping. If there were more objects in the scene, the sense of perspective between the various objects would be even keener. The entire code can be found under *Example10_1/Example10_1.cpp*.

In *Example10_2*, we use the glTransform routines to simulate camera motion. We define the key frames as before, but we actually will be using the rotation to simulate camera rotation as well. From frames 0 to 30, the camera is at the same position as in *Example10_1*.

```
SetKeyFrame(0,camKeyFrameValues, camKeyFramePositions, &camNumOfKeyFrames,
8.,7.,17.,1.,1.,1.,0.,0.,0.);
SetKeyFrame(30,camKeyFrameValues, camKeyFramePositions, &camNumOfKeyFrames,
8.,7.,17.,1.,1.,1.,0.,0.,0.);
```

From frames 30 to 50, we move the camera and rotate it slightly along the *x*- and *y*-axis to stage Snowy in the shot:

```
SetKeyFrame(50,camKeyFrameValues, camKeyFramePositions, &camNumOfKeyFrames,
13.,4,7.,1.,1.,1.,5.,10.,0.);
SetKeyFrame(70,camKeyFrameValues, camKeyFramePositions, &camNumOfKeyFrames,
13.,4,7.,1.,1.,1.,5.,10.,0.);
```

Finally, in the Display code, we interpolate the camera position values.

```
EvaluateXformsAt(camKeyFrameValues, camKeyFramePositions, camNumOfKeyFrames,
currentFrame, camXforms);
```

Then we apply the camera transforms using the gl transformation routines. Notice that we apply rotations first, since we want the camera to move about its local pivot point. The translations are applied in the opposite direction as specified:

```
glLoadIdentity();
glRotatef(camXforms[5], 0.,0.,1.);
glRotatef(camXforms[4], 0.,1.,0.);
glRotatef(camXforms[3], 1.,0.,0.);
glTranslatef(-camXforms[0],-camXforms[1],-camXforms[2]);
```

Watch the camera zoom in on Snowy. In general, this technique is easier for animating the camera, and we shall use it more in the next few sections.

10.2 Building up a real time 3D game

The snowman animation was a simple example of a camera animation. Animating the camera can attain even more interesting results. In games and motion rides, viewers are given the illusion of moving around in the 3D world from the perspective of their own eyes—that is, as if they are sitting on the camera and moving it along with them. Let us explore how we can achieve this result by building up a 3D game.

In *Example10_3*, we develop a game that entails a user exploring a 3D terrain. The user can move himself back and forth and reorient himself in this terrain. The user will be allowed to control the motion of the viewpoint by pressing on the arrow keys on the keyboard. (In a typical game setting, this functionality would be provided by a joystick.) In this example, we use the keyboard up-down keys to move the camera along the *x-z* plane and the left-right keys to enable the user to re-orient himself left and right.

We shall not be going into the mechanics of how we designed this game since detailed those points in Chapter 4. In general, you should follow that process when designing your own games.

Camera Definition

First, let us see how to define our camera. In the last example we used a nine-element array to hold the camera position. In this example, it will be easier to think of the camera in terms of its position and orientation exclusively: it is rare for someone to scale a camera in the middle of a game. We define a structure for the camera as follows:

```
typedef struct _tCamera
{
        GLfloat position[3];
        GLfloat orientation[3];
} CAMERAINFO;
```

We then define an actual camera:

```
CAMERAINFO camera;
```

We define the initial camera position to be at a default position

```
void InitCameraPosition(CAMERAINFO *camera){
        camera->position[0] =0;
        camera->position[1] = 0.2;
        camera->position[2] =0 ;

        camera->orientation[0] += 0.0;
        camera->orientation[1] = 0.0;
        camera->orientation[2] = 0.0;
}
```

that is, slightly above the ground and looking down the *z*-axis.
To apply motion to the camera, we simply apply the gl Transforms we saw earlier, using the current camera position:

```
void MoveCameraPosition(CAMERAINFO camera){
        glRotatef(camera.orientation[2], 0.,0.,1.);
        glRotatef(camera.orientation[1], 0.,1.,0.);
        glRotatef(camera.orientation[0], 1.,0.,0.);
        glTranslatef(-camera.position[0],-camera.position[1], -camera.position[2]);
}
```

The main question is this: How do we update our camera position based on user input in real time?

Camera Update

To update the camera position, we make use of the popular physics concept:

Distance traveled - velocity*elapsed time

Therefore:

Final Position - Initial Position + velocity*(elapsed time)

The camera position is updated at every tick of the game logic. The velocity of the motion is determined at every tick by checking the state of the arrow keys on the keyboard. If the up arrow is being pressed, then velocity is positive, causing forward motion. If the down key is being pressed, velocity is negative, causing backward motion. If nothing is being pressed, then velocity is 0, causing no motion to happen. The difference between the last timer tick and the current one gives the elapsed time.

The keyboard call-back function that we used earlier only kicks in when a state change occurs – a key is pressed or released. The new formula we are needs us to identify the state of the key at every *tick*. We make use of the Windows function:

```
SHORT GetAsyncKeyState(int vKey)
```

This function checks the state of the key defined by the parameter vKey—is it being pressed or not. Unix platforms provide a similar library call.
For simplicity, we define our own Boolean macro:

```
#define KEYDOWN(vkcode) (GetAsyncKeyState(vkcode) & 0x8000) ? 1 : 0
```

This function tests whether the key identified the specified *vkcode* is currently being pressed down and returns 1 if it is. For the up, down, left and right arrow keys, the *vkcode* ID is defined as: VK_UP and VK_DOWN, VK_LEFT and VK_RIGHT. Using this function, we can set up a test to determine key state and identify the velocity appropriately. Note that only one key is assumed to being pressed at a given tick.

```
if (KEYDOWN(VK_UP))
        velocity - 0.4f;
if (KEYDOWN(VK_DOWN))
        velocity - -0.4f;
```

Similarly, the orientation of the camera can be determined by using angular velocity. We use the math equation:

Final orientation of the camera = initial orientation + avelocity* elapsed time

The left and right arrow keys define the angluar velocity as follows

```
if (KEYDOWN(VK_LEFT))
        avelocity = -7;
if (KEYDOWN(VK_RIGHT))
        avelocity = 7;
```

High resolution timing is supported in Win32 by the **QueryPerformanceCounter()** and **QueryPerformanceFrequency()** API methods. The first call, **QueryPerformanceCounter()**, queries the actual value of the high-resolution performance counter at any point. The second function, **QueryPerformanceFrequency()**, will return the number of counts per second that the high-resolution counter performs. To retrieve the elapsed time from the previous tick, you have to get the actual value of the high-resolution performance counter immediately before and immediately after the section of code to be timed. The difference of these values would indicate the counts that elapsed while the code executed. The elapsed time can be computed by dividing this difference by the number of counts per second that the high-resolution counter performs (the frequency of the high-resolution timer).

```
elapsedTime = (GLfloat)(currentTime.QuadPart - lastTime.QuadPart)/frequency.QuadPart;
```

In this example we are driving the viewer along a terrain. The viewer cannot lift himself off the ground (motion along the *y*-axis). He also can only rotate his viewpoint about the *y*-axis (along the *x-z* plane, called a roll), but not along the *x*-axis (called a pitch) or *z*-axis (called a yaw). If you want to simulate a flying motion, you will have to enable this motion as well.

The distance traveled along the *x*- and or *z*-axis as a function of the total distance traveled depends on the orientation of the camera position as shown in the Fig.10.1.

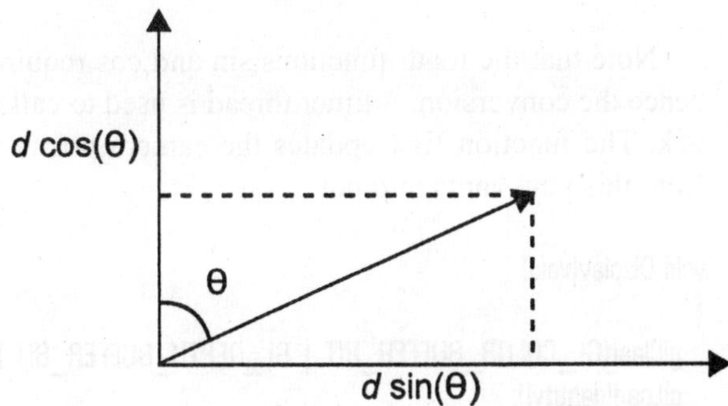

Fig.10.1: Distance traveled in the XZ plane

Using this, we can define the function to update the camera position:

```
void UpdateCameraPosition(CAMERAINFO *camera) {
```

```
            LARGE_INTEGER currentTime;
            GLfloat elapsedTime;
            GLfloat velocity=0;
            GLfloat avelocity= 0;
            Glfloat d = 0;

            // if up arrow has been pressed, velocity is now forward
            if (KEYDOWN(VK_UP))
                    velocity = 0.4f;
            if (KEYDOWN(VK_DOWN))
                    velocity = -0.4f;
            if (KEYDOWN(VK_LEFT))
                    avelocity = -7;
            if (KEYDOWN(VK_RIGHT))
                    avelocity = 7;

            QueryPerformanceCounter(&currentTime);
            elapsedTime = (GLfloat)(currentTime.QuadPart - lastTime.QuadPart)/frequency.QuadPart;
            lastTime = currentTime;


            camera->orientation[0] = 0.;
            camera->orientation[1] += avelocity*elapsedTime;
            camera->orientation[2] = 0.0;
            d = elapsedTime * velocity;
            camera->position[0] += d * ((GLfloat)sin(TORADIANS(camera->orientation[1])));
            camera->position[1] = 0.2;
            camera->position[2] -= d * ((GLfloat)(cos( TORADIANS(camera->orientation[1])));
}
```

Note that the math functions sin and cos require the angles to be in radians, hence the conversion. A timer thread is used to call the *Display* function at every tick. The function first updates the camera position, and then draws the world from this new vantage point.

```
void Display(void)
{
  glClear(GL_COLOR_BUFFER_BIT | GL_DEPTH_BUFFER_BIT );
  glLoadIdentity();

  UpdateCameraPosition(&camera);
  MoveCameraPosition(camera);

  drawWorld();
```

```
 glFlush();
 glutSwapBuffers();

}
void timer(int value)
{
 // Force a redisplay .. , also contains game logic
 glutPostRedisplay();
// Restart the timer
 glutTimerFunc(10, timer, 0);
}
```

Defining the World

Let us now see how we can define a world to view with our camera! For this viewpoint animation, we will make use of a textured ground as the land strip that we are driving through.

The Ground

The extents of the ground plane extend from -100 to 100, and it is defined as follows:

```
void draw_Ground()
{
        glBindTexture(GL_TEXTURE_2D, textureIDs[0]);
        glBegin(GL_QUADS);
        glNormal3f(0,1,0);
        glVertex3f(100.,0,-100);
        glVertex3f(-100.,0,-100);
        glVertex3f(-100.,0,00);
        glVertex3f(100.,0,100);
        glEnd();
}
```

So actually, we should clamp our camera motion to never go beyond the maximum boundaries in x and z of (-100,100).

In keeping with our outer-space theme, we assign the ground a texture of a bumpy surface, as shown in Fig.10.2. This image is actually an image of the Mars terrain that we downloaded from the Internet. You should be able to find tons of images on the Internet for your projects. Just remember that texture images needs to be a BMP file with a size that is a power of 2 such as 256, 512, or 1024 etc. You may notice that we did not use the *glTexCoord* function to define the mapping of the texture coordinates. Instead, we use OpenGL's functionality to automatically generate texture coordinates.

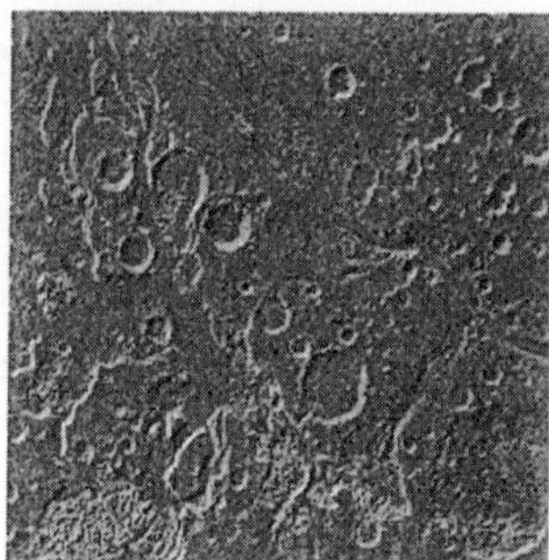

Fig.10.2: A mars terrain

Automatic Texture Generation

OpenGL has three automatic texture generation mode. One is used when doing spherical environment mapping. The other two are eye linear and object linear. In eye linear mode, a texture coordinate is generated from a vertex position by computing the distance of the vertex from a plane specified in eye coordinates. In object linear mode, a texture coordinate is generated from a vertex position by computing the distance of the vertex from a plane specified in object coordinates. We shall use object linear in this example.

We need to do two things:

- tell OpenGL that we will use the automatic texture generation with object linear. This is done by making a call to the function *glTexGeni()* with the parameter, GL_OBJECT_LINEAR
- tell OpenGL which plane we will use.

How to get the plane equation is tricky. Every plane is specified by four parameters and defined by the equation: P_1*x+ P_2*y+ P_3*z + P_4 = 0. Using GL_OBJECT_LINEAR, the texture coordinate at an vertex (x0,y0,z0,w0) is given by P_1*x0+ P_2*y0+ P_3*z0 + P_4.*w0 (w0 is 1 usually.)

The P_1, P_2, P_3 and P_4 values are supplied as arguments to glTexGen*() with *pname* set to GL_OBJECT_PLANE. With P_1, P_2, P_3 and P_4 correctly normalized, this function gives the distance from the vertex to a plane. For example, if $P_1 = P_2 = P_4 = 0$ and $P_3 = 1$, the function gives the distance between the vertex and the plane $z = 0$.

The distance is positive on one side of the plane, negative on the other, and zero if the vertex lies on the plane. Since our ground lies in the *x-z* plane, we want to generate s and t coordinates by the vertex distance from the *z*=0 plane and the *x*=0 plane.

The code to automatically generate OpenGL texture coordinates can be defined as:

```
// OpenGL's automatic generation of texture coordinates

        /* Use the Z-0 and X-0 planes for s and t generation because the ground lies
along the x-z plane. */
        float tPlane[4] - {0., 0., 1., 0.}; // This is the Z-0 plane
        glTexGenfv (GL_T, GL_OBJECT_PLANE, tPlane);

        float sPlane[4] - {1., 0., 0., 0.};
        glTexGenfv (GL_S, GL_OBJECT_PLANE, sPlane);


// Generate the texture coordinates based on object coords distance from the axis.
    glTexGeni (GL_S, GL_TEXTURE_GEN_MODE, GL_OBJECT_LINEAR);
    glTexGeni (GL_T, GL_TEXTURE_GEN_MODE, GL_OBJECT_LINEAR);

            // Enable OpenGL's automatic generation of texture coordinates
    glEnable (GL_TEXTURE_GEN_S);
    glEnable (GL_TEXTURE_GEN_T);
```

Refer to [HILL00] for more details on how to automatically generate texture coordinates.

But just driving through an empty terrain is boring! Let's throw some more models into the scene.

The Android

Let us define the Android we saw in earlier Chapters to be part of this world. Maybe the object of the game is to locate Andy—our favorite Android! We define Andy with a golden metallic material. The code to draw the Android at a given position defined in an array called *position* is shown below:

```
void draw_Andy(GLfloat *position){
            int i,j;
  glMaterialfv(GL_FRONT, GL_SPECULAR, mat_specular);
  glMaterialfv(GL_FRONT, GL_SHININESS, low_shininess);
  glMaterialfv(GL_FRONT, GL_AMBIENT, mat_ambient);
  glMaterialfv(GL_FRONT, GL_DIFFUSE, mat_diffuse);
  glPushMatrix();
  glTranslatef(position[0],position[1]+0.5,position[2]);
  glScalef(0.2,0.2,0.2);

            for (j-0;j<noofshapes;j++) {
                        for (i-0;i<noofpoly[j]*3;i-i+3) {
                                    glBegin(GL_TRIANGLES);
```

```
                        glNormal3fv(&(normals[j][3*nindices[j][i]]));
                        glVertex3fv(&(coords[j][3*indices[j][i]]));
                        glNormal3fv(&(normals[j][3*nindices[j][i+1]]));
                        glVertex3fv(&(coords[j][3*indices[j][i+1]]));
                        glNormal3fv(&(normals[j][3*nindices[j][i+2]]));
                        glVertex3fv(&(coords[j][3*indices[j][i+2]]));
                glEnd();
                }
        }
        glPopMatrix();
}
```

Some Clutter

We randomly throw in a few cylinders to clutter up the scene. A cylindrical model can be created by using the glu library command

gluCylinder (GLUquadric* quad , GLdouble base , GLdouble top , GLdouble height , GLint slices , GLint stacks);

quad: specifies the quadrics object
base: specifies the radius of the cylinder at z=0.
top: specifies the radius of the cylinder at z=height.
height: specifies the height of the cylinder.
slices: specifies the number of subdivisions around the z axis.
stacks: specifies the number of subdivisions along the z axis.

The quadric object need only be defined once using the function

GluNewQuadric.

In our game, the cylinders are drawn with a texture map wrapped around them, as shown in Fig.10.3. This map causes the cylinders to look like buildings with lit-up windows

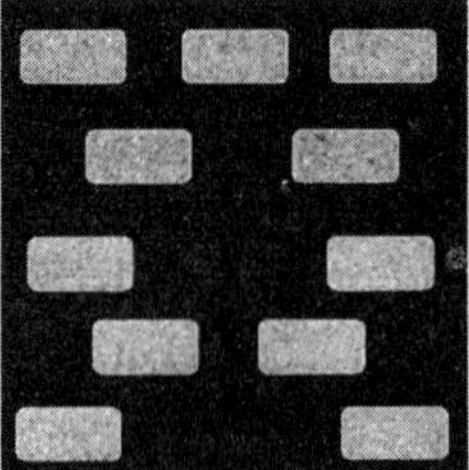

Fig.10.3: Texture map for the cylinders

```
void draw_Cylinder(GLfloat *position){
```

```
  glBindTexture(GL_TEXTURE_2D, textureIDs[1]);
  glPushMatrix();

  glTranslatef(position[0],position[1]+3,position[2]);
  glRotatef(90,1,0,0);
  gluCylinder( cylinder, 1.,1., 7., 30,30 );

  glPopMatrix();
  glBindTexture(GL_TEXTURE_2D, 0);
}
```

You can make the game a lot more interesting by defining more interesting and complicated models in the 3D world.

Putting the World Together

We define a simple structure to store the position and type of all the models in the scene.

```
typedef struct _tModels
{
        GLfloat position[3];
        GLint type;
} MODELINFO;
```

The two types of models we define for our game are

```
#define TANDY 1
#define TCYLINDER 2
```

The models are stored in an STL vector map of the *MODELINFO* struct.

```
vector<MODELINFO *> Models;
```

We now define a function called *InitWorld*, which randomly positions Andy within the extents of our world. It also creates the quadric object and the 25 cylinders in the world. The object of the game is to find Andy hidden amongst all this clutter.

```
void InitWorld(){
        int i;
        SYSTEMTIME systime;
        GetSystemTime(&systime);
        // Seed random number generator
        srand(systime.wMinute*60+systime.wSecond);
```

```
            noofshapes     =     ReadVRML("..\\Models\\robot.wrl",     &coords[0][0],
&normals[0][0],&indices[0][0],&nindices[0][0], &(noofpoly[0]), MAXSHAPES, MAXCOORDS);

        MODELINFO *Andy, *Cylinder[25];
        Andy = (MODELINFO *)malloc (sizeof(MODELINFO));
        Andy->position[0] = RandomPos();
        Andy->position[0] = 0;
        Andy->position[1] = 0;
        Andy->position[2] = RandomPos();
        Andy->position[2] = 0;
        Andy->type = TANDY;

        Models.push_back(Andy);
        cylinder = gluNewQuadric( );          /* Allocate quadric object */
        gluQuadricDrawStyle( cylinder, GLU_FILL ); /* Render it as solid */
        for (i=0;i<25;i++){
                Cylinder[i] = (MODELINFO *)malloc (sizeof(MODELINFO));
                Cylinder[i]->position[0] = RandomPos();
                Cylinder[i]->position[1] = 0;
                Cylinder[i]->position[2] = RandomPos();
                Cylinder[i]->type = TCYLINDER;
                Models.push_back(Cylinder[i]);
        }
```

The code for the *drawWorld* function is shown below. It simply draws the ground, and then loops through all the models defined in the Models vector, calling the appropriate drawing routines.

```
void drawWorld(){
        vector<MODELINFO *>::iterator it;
        glTexEnvf(GL_TEXTURE_ENV, GL_TEXTURE_ENV_MODE, GL_DECAL);

        draw_Ground();

        for (it = Models.begin(); it!= Models.end(); ++it) {
                if ((*it)->type == TANDY)
                        draw_Andy((*it)->position);
                else if ((*it)->type == TCYLINDER)
                        draw_Cylinder((*it)->position);
        }
}
```

The entire code can be found under *Example10_3*, in files: *Example10_3.cpp, Example10_3.h, Models.h and Models.cpp*. You will also need to include the provided files, *bmp cpp*, *bmp.h*, *vrml.cpp* and *vrml.h* to compile and link this project.

Run the program, and travel through your world. Can you find Andy? You will find that you can actually travel through the objects! In an actual game, we would clamp the camera's motion so that we would not be able to do so.

Try adding more cylinders and other models to your program. As you add more and more objects, you may also notice a serious slowdown in speed. This is because there are so many objects in our scene! One serious drawback in OpenGL is that OpenGL does not understand the notion of objects; it just knows about the polygons that make up the models. So it actually tests all the polygons of every model to check whether they should be drawn or clipped. Each cylinder model itself has about 100 polygons. Making 100 calls per cylinder can degrade performance a lot. Instead, if we could perform a simple check and not draw objects (cull objects) that are out of view, we could improve performance substantially.

Object Culling

Remember that our camera is defined as a frustum with a clipping range of about 50 units:

```
gluPerspective(35., (GLfloat)w/h, 0.5, 50.0);
```

The easiest way to check whether (entire) objects are within the camera view is to approximate the viewing range with a sphere. We can define the sphere that encompasses the viewing frustum as a sphere located at the approximate center of this frustum

```
sphCenter[0] = camera.position[0] + sin(TORADIANS(camera.orientation[1]))*25.;
sphCenter[2] = -camera.position[2] - cos(TORADIANS(camera.orientation[1]))*25.;
```

with a radius of 50 units. Any object that is not within the boundaries of this sphere will be eliminated. Recall from math that the distance of point (x,z) from a center (cx,cy) can be computed as

```
sqrt((x-cx)(z-cz))
```

If this distance is less than the radius of our sphere, then we are within the viewing distance (or close to it); otherwise we are outside. We can use this test to determine whether the object is positioned within this sphere or not. The code

```
bool WithinViewingSphere(GLfloat *sphCenter, GLfloat *pos){
	if ( ((sphCenter[0]-pos[0])*(sphCenter[0]-pos[0])
			+ (sphCenter[2] - pos[2])*(sphCenter[2]-pos[2])) < (50.*50.))
			return TRUE;
```

```
        return FALSE;
    }
```

returns *TRUE* if an object located at a position defined by the array *pos* is inside the viewing sphere.

The drawWorld code is now modified to test if objects are within the viewing sphere. If not, we do not make the openGL call to draw them, effectively culling them from the scene.

```
for (it = Models.begin(); it!= Models.end(); ++it) {
                if (!(WithinViewingSphere(sphCenter, (*it)->position)))
                        continue;
                else {
// draw the objects
```

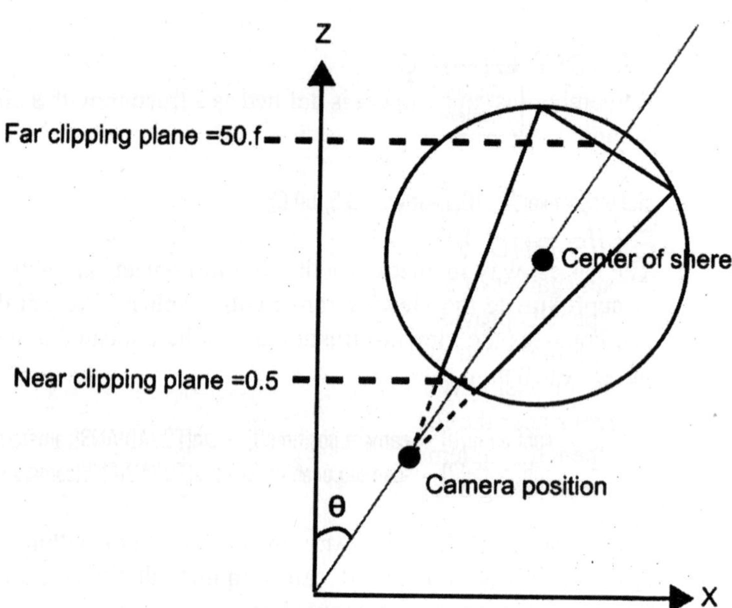

Fig.10.4: Sphere approximating the viewing frustum

The structure holding the models is often referred to as the *scene graph*. Usually the graph is structured as a tree. Real objects occupying a certain region of space in the scene are placed at leaf nodes. Nearby children are grouped together under a parent node. The parent node (usually a null node) defines a bounding region that encompasses the area occupied by all the children underneath. These parents can be further grouped together and so on. The top-level parent occupies the entire extent of the world. This structure helps tremendously in fast culling. If you reach a node that is not in the scene, then there is no need to traverse to the objects beneath this node. All the objects below this node will be culled. For example, in Fig.10.5, ,if we reach nodeA and find that its extents are not within view, then Sphere1 and Sphere2 are automatically culled out as well. The test to check for culling is similar to what we discussed above.

Many times, culling is not just a simple on-off switch. You can define different levels of detail for the model. As the model gets closer to the viewpoint, the more detailed version of the model is used.

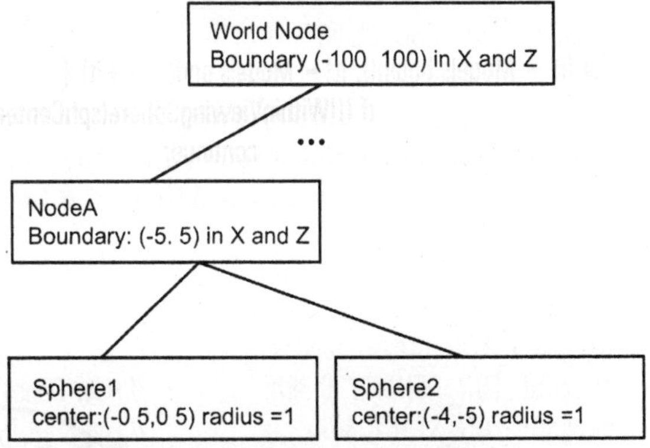

Fig.10.5: Scene Tree

Bells and Whistles: Adding Fog

In many games of today, fog and mist are added to simulate a more moody scene.

OpenGL provides a primitive capability for rendering atmospheric effects such as fog, mist, and haze. It is useful to simulate the impact of atmospheric effects on visibility to increase realism. It also allows the designer to cull out polygons near the fog limit since they cannot be seen anyway.

OpenGL implements fogging by blending the fog color with the incoming fragments using a fog blending factor, *f*,

$$C = fC_{in} + ((1 - f)C_{fog}$$

The function

```
glFogi(GL_FOG_MODE, fogfilter);
```

establishes the fog filter mode. It can be of type GL_EXP, GL_EXP2 or GL_LINEAR. GL_LINEAR is the best for fog-rendering mode; objects fade in and out of the fog much better.
The command

```
glFogfv(GL_FOG_COLOR, fogcolor);
```

sets the color of the fog.

The command

```
glFogf(GL_FOG_START, 1.0f);
```

will establish how close to the screen the fog should start. You can change the number to whatever you want depending on where you want the fog to start. The command

```
glFogf(GL_FOG_END, 25.0f)
```

This tells the OpenGL program how far into the screen the fog should go. Objects farther than this distance from the camera cannot be seen and hence can be culled out. The code below defines fog for our game using OpenGL.

```
GLfloat fogColor[4]= {0.5f, 0.5f, 0.5f, 1.0f};      // Fog Color
glFogi(GL_FOG_MODE, GL_LINEAR);                     // Fog Mode
glFogfv(GL_FOG_COLOR, fogColor);                    // Set Fog Color
glFogf(GL_FOG_DENSITY, 0.35f);                 // How Dense Will The Fog Be
glFogf(GL_FOG_START, 1.0f);                    // Fog Start Depth
glFogf(GL_FOG_END, 25.0f);                          // Fog End Depth
glEnable(GL_FOG);
```

Run your program with fog enabled and watch what it does to enhance the mood of your game. An image from our game is shown in Color Plate 6.

There are many more bells and whistles you can add. You can have rooms and dungeons set up for the viewer to explore. Andy himself may not be stationary—he may be animating about the scene, trying to avoid you. Or, he may be firing bullets at you! You have to hide behind an object, locate him and fire back. If you hit him first, you win! Otherwise...

Summary

In this chapter we have explored how to use camera animation within a 3D world. We animated the camera in the snowman animation to achieve a smooth camera pan to zoom into snowy. We have also seen how to use camera motion to simulate a viewer moving through a 3D world. We captured keyboard input to define the motion of the camera to develop a cool 3D game.

There is no end to the creativity you can put into your graphics. Explore some ideas yourself and see how you can implement them.

Chapter **11**

Lights, Camera, Action!

In the last few chapters, we looked into the principles of animation and how to bring our models to life. In this chapter, we shall put together all that we have learned to develop our own animated movie using Maya PLE.

We shall go through the development of the movie, just as the pros do-from pre-production, where the story is developed; to production, where the movie is developed; to post-production, where the frames are polished up and finally put onto tape..

By the end of this chapter, you will know all the industry buzz words and secrets!

In this chapter, you will learn

- Steps taken to develop a movie-from pre-production to post-production
- Animation using Maya
- Create a fully rendered movie sequence using Maya

Let us start the production of our movie.

11.1 Pre-production

We are ready to make a movie, we have the technical know-how-but wait! What is the story? As a famous director put it, "Without the story, you have nothing." No amount of technology can save a bad story. The story is paramount to the success of a project.

Pre-production is the first step in creating an animated project. It's the step where the story is identified and storyboarded. How does one get ideas for a story? As you read through this section, your brain will be buzzing with story ideas-good ideas are really not that hard to come by. (Great ideas, of course, are not that easy, and take a lot of time and perseverance).

The Story: Idea is Everything

The story is where you identify the main idea of your movie. It's where you identify the conflict, the resolution and the punch line of the production. Most films that really grab you can be summed up in one line, called the pitch:

- ***Jaws:*** "Man afraid of water pursues killer shark": horror movie for adults
- ***Toy Story:*** "Toys come to life": humorous movie for kids
- ***A Fish Story:*** "A fish is thirsty, but water is not enough": humorous movie for readers of this book.

Try to create a couple of these pitches by asking yourself "What if...?"

For example: "What if we had super power?" (*Incredibles*), "What if..toys came to life?" (*Toy Story*) or "What if a fish wanted to taste martini?" (*our movie*). Write down as many as you can as fast as you can. Scribble a couple of notes down around them. At this point don't self-censor, write everything down however dumb it seems. It's part of the creative process called brainstorming.

Ideas come from everywhere. Sitting in an empty room isn't going to inspire you-read a daily newspaper, get out to the theatre or nip down the pub. Don't push trying to get ideas; they'll come. Consider keeping some sort of journal so you can safeguard everything from getting lost.

Once you have identified an idea, you need to expand it into a full-fledged story. The story needs to be visual (so no introspective characters) and, like any good story, has to have a beginning, middle and an end. You have to fully exploit your idea, so if you've got a great beginning, make something else happen in the middle and then give it a good resolution. The story and the characters have to go somewhere.

Boil a story down to its basics and what usually happens is this: A character gets involved in some sort of situation that gives the character an aim. The story relates how the character works to achieve this aim. Just before the end comes the "make-or-break" time and the aim is usually achieved in some climactic finale.

Audience

Just as we learned in Chapter 4, the audience is paramount. Detailed research goes into identifying the audience for a film and their likes and dislikes.

The characters in the story go a long way toward helping the audience identify with the film. The character' struggles and aims should be something with which the audience can relate. For example, consider the characters in *Toy Story*-the toys in the movie are commonplace toys that everyone can relate having playing with. And which kids hasn't wondered if their toys could come to life.

The great thing about the pre-production stage is that you can talk through your ideas with others. Tell your story to your target audience. The response you get-"Why doesn't this happen?", "Couldn't she be a suspect?", "I'm not sure about the ending, but if they did escape..."-will help your story evolve. You will find yourself thinking out loud, adding new parts, cutting back on stuff that gets a bad reception. It's the tradition of oral storytelling: audience response is critical in refining your plot.

In the end, it all comes down to two things:

1) Do you love the idea? You'd better, because you'll be making the movie.
2) Will your audience love the idea and the characters in the story?

Make a movie people want to watch.

For this book, let us design a movie called *A Fish Story*. The audience is you, our readers. We do love the story, and hope you will love it as well.

The story line is as follows:

A fish is swimming in a fish bowl. Adjacent to the fish bowl is a martini in a glass. The fish decides it's thirsty, and it wants martini, not water! It jumps over to the martini glass: only to find the glass is filled with water too!

Storyboards

We saw storyboards in Chapter 4. Let us explore them in more detail in this chapter. The concept of storyboarding is rather simple. Storyboarding is a way for you to organize and preplan your project before attempting to actually develop it. The storyboarding phase helps identify the shots of the story in greater detail. It identifies the key frames of the shot, the layout of the characters, and the general ambient setting and mood of the scene. It also identifies the rough timing of the shots.

The storyboard is divided into shots. Each shot has key frames that are drawn out to represent the essence of that shot. The task is to move your story, from frame to frame, filling in the details.

You don't have to be an artist to storyboard. If you have ever looked at storyboards by Hitchcock or Spielberg, you have to admit that they can't draw. Storyboarding is especially useful for complex visual sequences e.g. elaborate shots or special effects sequences. A detailed timing chart is also made at this point.

For our fish story, we identify five shots as described below. The story boards for the shots are also shown. The timing is calculated based on the way the story boards play out. It takes a lot of experience to predict good timing. Trial and error is part of this process.

- Shot1: Opening shot: the camera pans in to show the fish swimming in a bowl on a table. A martini glass sits next to the bowl. Shot1 ends with a close-up of the fish wiggling his eyes. He has an idea! We define Shot1 to be 70 frames long - just a little more than 2 seconds.

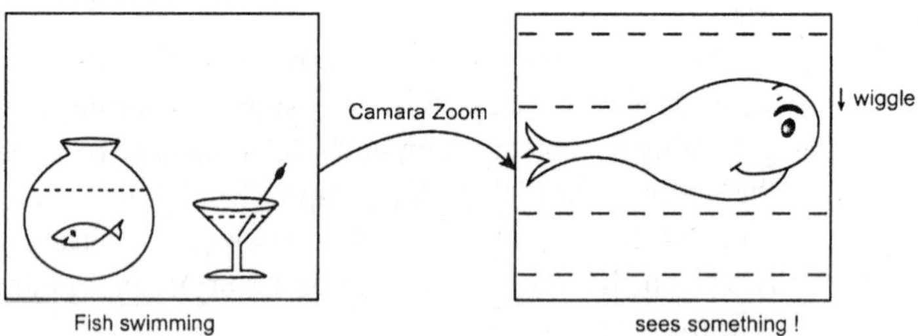

Fig.11.1: Camera pan and zoom on to fish

- Shot2: Cut to show the martini glass. This was what captured the fish's attention. Since Shot2 is just a still of the martini glass, we define it to be about 20 frames only.

Fig.11.2: Cut to the martini glass

- Shot3: Camera is back to the fish. The fish raises his eyebrows to indicate he sees the glass. He gets ready to leap over to the martini glass. Shot3 is about 50 frames long.

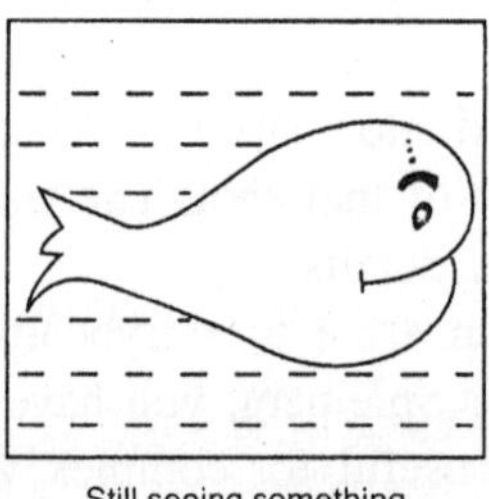

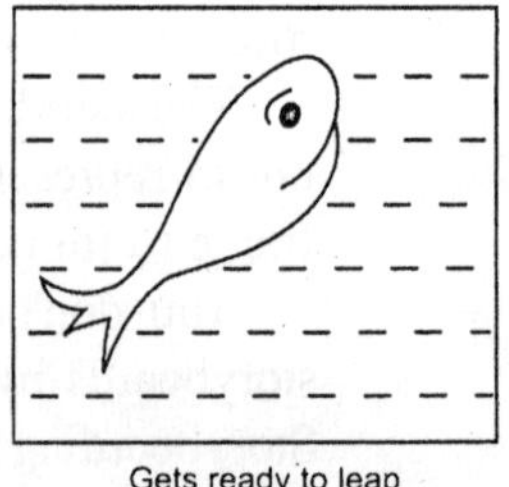

Fig.11.3: cut back to fish who is getting ready to leap

■ Shot4: Camera pans back. The shot shows the fish leaping and reaching the martini glass. Shot4 is 60 frames long.

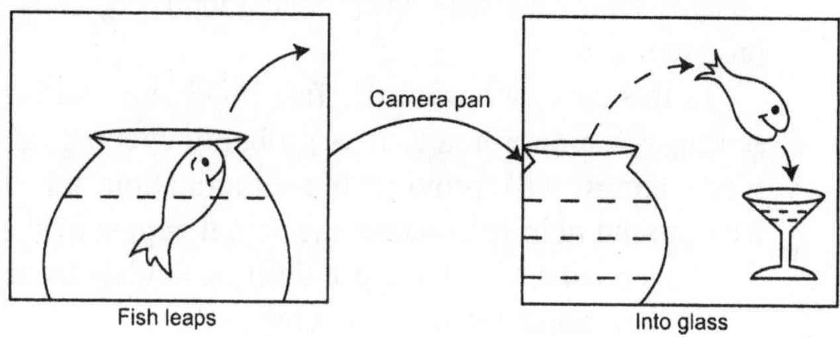

Fig.11.4: Fish leaps

■ Shot5: Camera closes in on the fish in the martini glass. Oh no! It's not a martini; it's only water! Shot5 lasts about 50 frames.

Oh no ! Only water

Fig.11.5: Its only water !

Here are a few tips for producing quick-and-dirty storyboards.

1) **Keep the area you have to draw small.** This allows you to draw much faster. The pictures become more like doodles than works of art.

Remember, the point is to get an idea of how things will look on screen.

2) **Copy a set of storyboard sheets** so you don't have to spend all night drawing screen boxes.
3) **Sketch in pencil** so you can make changes easily. Then ink in for photocopying. Feel free to use any medium you are happy with professional storyboard artists use everything from magic markers to charcoal.

4. **Scribble down short notes** about what's happening in the shot (e.g., "Bob enters") what the characters are saying ("Is this it? Is this how...") or what kind of sound effects you want ("Roll of thunder").
5. **Number your shots** so that they can be quickly referenced to on the shot list and during editing.

Look and Feel

Based on the storyboards, a more detailed environment is defined. The mood of the scene is identified to guide the color choices and general lighting schemes.

Each character is defined in more detail, especially its general size and proportions in relationship to the other characters. Once the story boards and the general look and feel have been identified, it is time to actually create the production.

In this case, we want the fish bowl and martini glass to be on a table. The setting is inside a home. It is probably evening, since the martini glass is out-room lamps will provide the illumination for the scene. A mirror in the background help to enhance the visual appeal of the room.

(Its amazing - we have pre-defined models for the above scene from Chapter 8. We shall reuse them in this chapter.)

11.2 Production

Production is the phase of the project where the actual movie content is developed. In the CG world, production means: the modeling of the characters and sets, laying out the shots (composing the world), animating the models within the shots, and finally lighting the scene to achieve the desired ambience. In the end, you render the shots in order to see your animation in it's full glory. Start up Maya. Let's get ready to rumble.

Modeling

Based on the storyboards and character design, the characters and sets of the shot are modeled. The models can be input into the computer in different ways. Some models are first sculpted in clay and then input into the computer using 3D scanners. Some models are directly modeled in the computer using software like Maya.

Materials can also be assigned to the models at this point.

We already have most of the required models (sized appropriately) from Chapter 8.

We have created the fish and fish bowl model which you can import in: *Models/bowl.mb* and *Models/fish.mb*. Interested readers should try and model their own versions of the fish character.

Layout

The layout phase involves putting together the models to compose the 3D world. We saw how to lay out some of the world in Chapter 8. Let us put the fish and fishbowl into the scene as well.

1) Start up Maya.
2) Load in the world you created in Chapter 8. If you don't have one, load in our model from *Models/world.mb*.
3) Remove the second martini glass that we had created-we need to make room for the bowl!
4) Import bowl.mb.
5) Set selection mode to be *hierarchy* and select the bowl group.

6) Translate the bowl to (-3.7 ,3.1, -0.46) and set its scale to be (0.7, 0.7,0.7).
7) Set the selection mode to be *Object.*
8) If not already assigned, select only the outside bowl and assign it the *glass* shader that already exists, using the marking menu.
9) Assign the water inside the bowl to the water shader.
10) Import *fish.mb*. We have already assigned materials to the fish.
11) Select the entire fish group and translate it to (-2, 2.75,0).
12) Save your scene in a file called *worldwithfish.*

Animation

Now comes the fun part! Now we get to actually make each shot of our movie come alive.

Most animation systems (Maya included) use the frame as the basic unit of measurement for animation sequences. Each frame is played back in rapid succession to provide the illusion of motion. The frame rate (frames per second or fps) that is used to play back an animation is based on the medium in which the animation will be played back (for example, film, TV, video game, etc.) We saw that film requires 24 fps to depict smooth motion, whereas a TV requires 30 fps.

When you set a key frame (referred to as just key in Maya), you assign a value to an object's attribute: for example, it's translation, rotation, scale, color, etc., at the specified frame. When you set several keys for an object with different values, Maya will interpolate the in-between frames as it plays back the scene. The result is the movement or change over time of those objects and attributes-animation!

Let's setup Maya for animation.

1) Select the Animation menu set so that the animation related menu appears.
2) Select *Window > Settings/Preferences > Preferences* from the menu bar In the Preferences window that pops up, click the Settings category and set the *Time* option to *NTSC(30 fps)* so your animation will play at the rate of 30 frames per second. Click the Save button.

At the bottom of the Maya UI is the animation timeline, called the Time Slider in Maya lingo. Look over the playback controls, as shown in the figure below:

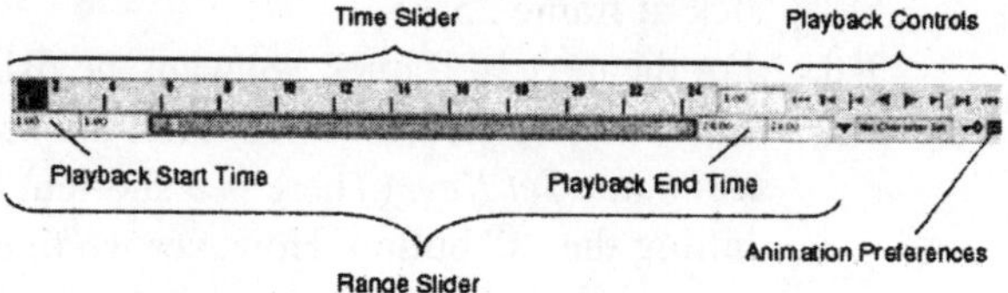

The Time Slider displays the frames of the animation along a time line. The key frames that you set for the selected object are displayed as red lines along the time line.

The Range Slider controls the range of frames that play when you click the play button. It displays the start time and end time of the entire animation, as well as Playback Start Time and Playback End Time in text boxes. The playback times may vary depending on which shot you want to view.

The box at the top right of the Time Slider lets you set the current frame (time) of the animation.

The Playback Controls control animation playback. You may recognize the conventional buttons for play and rewind (return to the start time). The stop button appears only when the animation is playing. To find out which operation a button/text-box represents, hold the mouse pointer over it.

The Animation Preferences button displays a window for setting animation preference settings such as the playback speed.

The above items will have more relevance as you work through this lesson.

In the *End Time* box, set the End Time to be 250. This sets the length of the entire animation to be 250 frames long.

Shot1

First, we will set up shot1, which is 70 frames long. In this shot, we will animate the fish. For 60 frames, Mr. Fish will swim till the end of the bowl, turn around and swim back up again. In the next 10 frames, Mr. Fish raises his eyebrows to indicate he is eyeing something interesting.

1. Click the rewind button to go to the start of the playback range. This changes the current frame to 0 (some systems use 1 as the start frame).
2. Make sure the selection mode is set to *hierarchy*.
3. Select the Fish.
4. Choose *Animate > Set Key* from the menu bar. This sets up the first key frame for the fish at the position defined. Notice the red marker that appears in the Time Slider at frame 0, known as a tick. Ticks in the Time Slider indicate the key frames for the currently selected object.
5. At frame 25, we want Mr. Fish to reach the end of the bowl. Go to frame 25. A convenient way to do this is to click at the desired frame in the Time Slider, or to enter this value in the Current Time box.
6. With the fish selected, go to the Channels Editor and set it's TranslateX=-5.
7. Select *Animate>Set Key* to define this keyframe. Make sure you see the tick at frame 25.
8. For the next 10 frames, we want the fish to turn around. Go to frame 35. Set the fish's RotateY = 180. Set the key frame by selecting *Animate>Set Key*. (There is a shortcut to setting the key frame-just hitting the "s" button. However we find that this doesn't always work very well.)
9. At frame 60, the fish reaches the other end of the bowl. Go to frame 60. Set the fish's TranslateX = -2 and set the key frame.
10. Play your animation by clicking the play button from the

Playback Controls.

From the three keys you've set, Maya creates motion between the positions. By default, the animation plays in a loop from frame 0 to 60. The animation occurs in the active window. If you have a fast computer, you might notice that the animation plays too fast. By default, Maya plays the animation as fast as it can be processed. Because this scene is simple, the animation might play faster than the default rate (30 frames per second). When you render all the frames of your animation for final production, the animation will be smooth. You can also drag the mouse back and forth (scrub) in the Time Slider to see the in between frames as you drag the mouse. This is the most basic aspect of our motion. Following the principles of animation, we can now enhance the animation. First, let's put in some secondary motion to make the movement look more believable.

Secondary Motion

The tail of the fish should swish as it swims around.

1. Rewind the animation.
2. Set the selection mode to *Object*.
3. Pick only the tail of the fish (you can pick it from the Windows>Outliner window as well, the object is called Tail and is under the Fish parent object.)
4. At frame 0, set the Tail to have a RotateY=20. This will swish the tail to one side. Set the key frame.
5. At frame 10, set the RotateY to -20. This will swish the tail in the other direction. Set the key frame.
6. Repeat steps 4-5 for frames 20,30,40,50 and 60. This will cause the tail to swish from side to side when you playback the animation.

Our shot now calls for a camera pan to zoom into Fishes expression - what we call staging. To do this, we actually need to create a new camera that can be animated. (Maya does not allow the default cameras to be animated.)

Staging

1. Choose *Create>Cameras>Camera* from the main menu.
2. A new camera will appear.
3. Open the Attributes Editor. In the Attributes Editor, you will see the transformation attributes as well as a tab section to define the camera

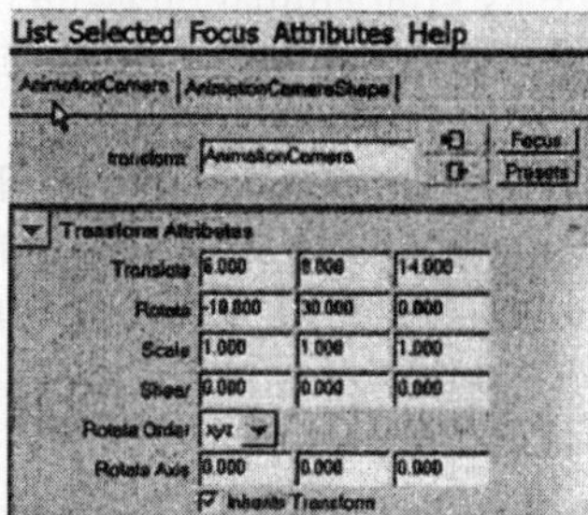

settings of this camera.

4. Rename the camera as *AnimationCamera* by clicking inside the box with the camera name.
5. Set the transforms of this camera to be Translate =(6., 8, 14), Rotation=(-10, 30, 0).
6. Choose the four panel layout (if it is not already chosen).
7. We don't need the side view window, so we will replace it with the view seen by our Animation camera.
8. In the menu bar of the side-view window, choose *Panels>Look through Selected.* Since the AnimationCamera was selected, this window will now show the world from the viewpoint of the animation camera.
9. Click into this window and hit 5 to view the objects in a shaded mode.

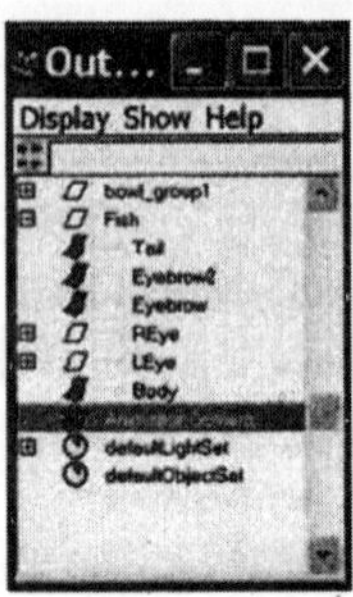

10. Play the animation with the new camera view selected.

This will be the camera we use for our movie. The animation of a camera is similar to animating objects. We need to define its key frames, and Maya will perform the interpolation for us.

We want the camera to be stationary for the first 30 frames. For this, we need to set a key frame at frame 30. From frames 30 to 60, we will let the camera zoom into Mr. Fish.

11. Go to frame 30.
12. Make sure the transforms of the Animation camera are still defined as Translate =(6, 8, 14), Rotate=(-10, 30, 0).
13. Make sure that the animation camera is the selected object, and hit *Animate>Set Key* to define this key frame.
14. At frame 60, define the key frame for the animation camera to have Translate= (-0.883, 3, 2). Keep everything else the same.

Now play your animation (with the animation camera window selected). You will see the camera zoom on the fish.

More Secondary Animation

From frames 60 to 70, we wish to show that the fish has seen something, and is conveying this to the audience by lifting his eyebrows in delight.

1. Go to frame 60. Select the eyeball of the Right eye (*REye*), by locating it in the outliner window.
2. Show the *Channels Editor*.

3. Set the Translate for the eyeball to be (-0.55, 0, 0.11) so that the fish appears to be looking in front.. Set the key frame.
4. The fish has seen something. He now looks at us with raised eyebrows.
5. At frame 70, move the eye ball to look at us by setting a key frame with the Translate values =(-0.56, -0.015, 0.11).

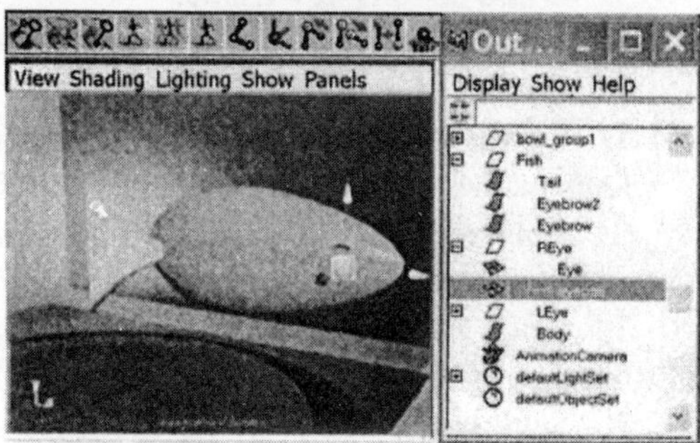

6. Pick *Eyebrow2* from the Outliner Window. Go to frame 60. Make sure the transforms are set as Translate=(0.063 , 0 , -0.109) and Rotate=(0, 16.866,0). Save this default position as a key frame.
7. At frame 70, raise the eyebrow by setting Ty = 0.05. Leave the other transform as is and set the key frame.
8. Select the *AnimationCamera.* Set frame 70 as a key frame with its current transformations. This will ensure that the camera stays still till frame 70.

Play your animation. You will see the fish looking up in delight. What has he seen? You can enhance the eye motion even further by making the eyebrow move up and down a couple of times and by widening the eye. We leave this as an exercise for the reader.

Shot2

Shot2 is a cut to the martini glass. It extends for 20 frames. Its main purpose in the film is to make the viewer aware of the fact that it was the glass that the fish saw.

1. At frame 71, we will cut to the martini glass. To do this, set the transforms for the camera as Translations=(1.06, 2.436, 5.382) and Rotations=(0, -30, 0) and save the key frame.
2. Go to frame 90 and save the key frame again to keep the camera stationary between frames 71 and 90.

Shot3

In shot3, we cut back to the fish. For continuity reasons, the camera position is the same that we ended at in shot1. Shot3 shows the fish excited and getting ready to launch out of the fish bowl. It extends from frame 91 to frame 140.

1. At frame 91, we cut back to the fish. Go to frame 91. Set the animation camera transforms back to Translation = (-0.883, 3, 2) and Rotation =(-10,

30,0). Save the key frame.
2. Go to frame 140 and set this as a key frame with the same transforms to keep the camera stationary during the shot.

The fish saw it! His eyebrows are wiggling.

3. Pick Eyebrow2. At frame 91, lower the eyebrow by setting Translate Y = 0 and set this key frame.
4. At Frame 94, lift the eyebrow by setting TranslateY = 0.05. Save the key frame.
5. Lower, lift and lower the eyebrows at frames 97, 100, and 104.

Let's get Mr. Fish to squash and stretch his body in anticipation of a leap.

6. Change the selection mode to *hierarchy* and pick the fish (make sure the entire hierarchy is chosen).
7. Go to frame 111 and save the fish's current transformation as a key frame.
8. Set a key frame at frame 130, by setting its Rotate=(0, 180 50), and Scale = (0.7, 1.2,1). Do not change the other transforms. This will squash the fish.
9. At frame 140, set a key frame by changing the fish's Scale=(1.2,0.8,1). This will stretch the fish in anticipation of the leap.
10. Play back the shot to view your animation. Again, you are encouraged to use your creativity in enhancing the motion to make it more believable.

Shot4

The fish is ready to fly. He wants some of that martini and he wants it now! In this shot, we shall pan back the camera so we can see the entire flight of the fish from bowl to glass. The shot lasts from frame 141 to frame 200.

1. At frame 141, we want to pan back to view the entire scene and see the fish leaping.
2. Set frame 141 as a key frame for the animation camera, with its transformations set as Translate= (6,6,14) and Rotation=(-10,30,0).
3. Go to frame 200, and save the current settings of the camera as a key frame as well.

Let's make the fish fly!

4. Pick the fish. At frame 141, set a key frame with the fish set at Translate = (-4.5,2.75,0), Rotate = (0,180,70) and Scale = (1.2,0.8,1)
5. At frame 170, set the key frame for the fish as Translate = (0.5,8,0), Rotate = (0,180,0) and Scale = (1,1,1).
6. At frame 200, set the keyframe for the fish as Translate = (3,5,0), Rotate = (0,180,-60) and Scale = (1.2,0.8,1). Almost in the martini glass!

Shot5

Shot5 is our finale. It shows the fish landing in the glass. But did he really get what he wanted?

For continuity reasons, we start the shot with the fish just about to land in the glass. Shot5 lasts from frame 201 to frame 250.

1. At frame 201, select the fish, and set its transforms as Translate = (3.1, 4.45, 0.47), Rotate = (0,180,-50) and Scale = (0.6,0.,6,0.6). The fish is about to land in the glass. The fish is a bit bigger than the glass, so we have to cheat on its size a bit! Set the key frame.
2. At frame 210, the fish has landed in the glass. Set frame 210 as a key frame with the fish transforms as Translate = (3.44,3.466,0.476) and Rotate = (-60,120,-50).
3. Oh no! It's only water!
4. Till frame 250, you may want to swish the fishs tail slowly, so the scene doesn't look completely still.

You can add in further animation detail to display the sadness that the fish is probably feeling at the end result of his effort. His eyebrows drooping or his eyes going half closed are great ways to convey a mood of sadness. What other motion can you think of?

Previewing the animation

To play a preview of the movie in 30 fps, we will use the *Playblast option* provided by Maya. Playblast enables you to preview the current animation. When you run Playblast on an animtion, the Playblast outputs a series of individual frames into a movie file format such as .avi (Windows), .iff (IRIX and Linux), and .qt (Mac OS X). This is an option is used for preview only!

From the menu bar, choose *Window>PlayBlast>*❐

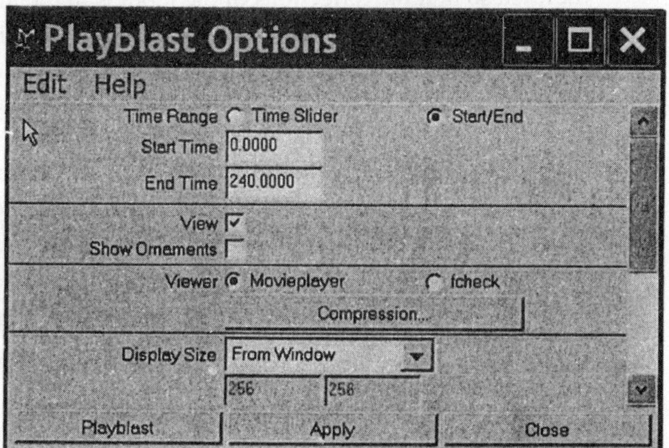

By default, Playblast previews the animation using the active view and the current time range in the Time Slider. In the Playblast window, set the Start/End option to be on, and change the *End Time* to be 140. This will enable us to preview shot1 to shot3 of the animation.

The default scale is 0.5, which makes the Playblast image resolution one-quarter the size of the active view. If the movieplayer option is on, Playblast uses the desktop video playback application specified in your Windows environment (QuickTime or RealPlayer) to view the Playblasted images. If you don't have a movieplayer, select the fcheck option. This will use Maya's internal movie playback format.

Click on *Playblast.* This will create a movie for you at the appropriate speed. Watch your movie go! Change the Start/End time appropriately to view shots 4 and 5. (Note: We have split up the shots because of size restrictions on our playback system. Your system may be able to handle the entire 250 frames.)

You can go back and change your animation if you don't like anything that you see.

Lights

Lights are used for much more than just illumination of the scene. The lighting in a shot can frame the characters, punch up the main idea/models of the shot, and define the mood of the scene. Wow!..and you thought lights were an afterthought.

In Chapter 8, we defined certain mood lighting that matches what we wish to portray here. We shall keep much of that same lighting. We add in a couple more lights to punch up the fish, and the martini glass.

To punch up the main characters, we define three spotlights.The first illuminates the fish bowl and the martini glass from the front. A top view of the layout is shown in the Fig.11.6

Two spotlights illuminate the models from the back. The back lighting is defined so that the reflections of the models in the mirror are illuminated as well. All three lights have a soft white color with an intensity of 0.5.

It's a good idea to render a few frames from every shot and make sure they are rendering as desired. Usually, every shot is fine tuned for lighting and material settings to make sure it looks just right!

We cheated on the light coming from the lampshade, so that our shadows

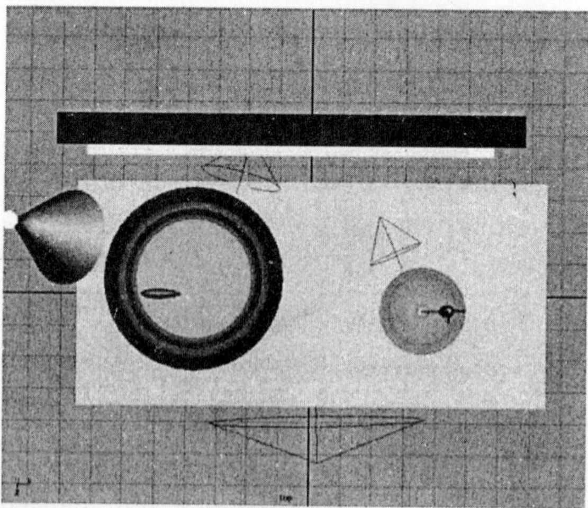

Fig.11.6: The top view

came out exactly where we wanted. Another point we noted while rendering our scenes: the fish was casting shadows in the glass bowl as well as in the water. This effect was not very pleasing, so we turned off the ability for the glass and

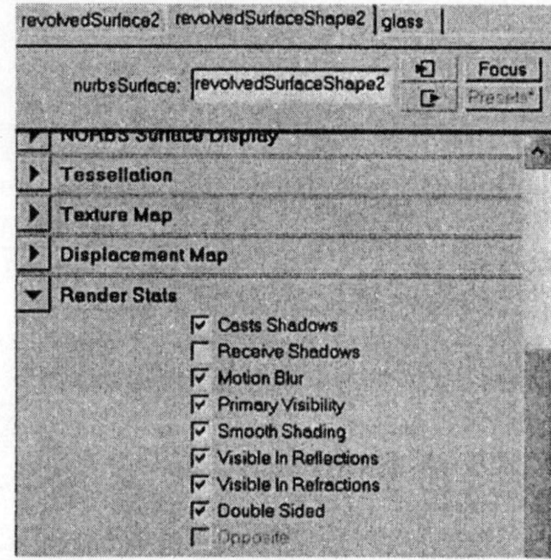

water materials to be able to receive shadows. You can do this by selecting the object, going to the Attributes Menu and turning off the radio button for *Receive Shadows* under the *Render Stats* option. Once you are satisfied with the test renders, we can set the frames off into batch render mode.

Render

Finally!! It is time to render all those images. Video resolution requires images be rendered with a resolution of 640 by 480. Film requires 1K resolution or higher.

We shall render our images to be 640 by 480. Maya has a batch render mode that allows you to render the entire animation sequence.

To set up the rendering job, proceed as follows

1. From the shelf, select the *Render Globals Window* icon to display the Render Global Settings window. In the Render Global Settings window, select the Common tab, and then open the *Image File Output section.*
2. In the Image File Output section, set the following options:
   a. File Name Prefix: Enter the name *fish.* This name will be the base of the filenames created by batch rendering. On top of the Render Globals window, you will see the Path: variable. This defines the exact folder in which the rendered files will be placed. Make a note of it.
   b. Frame/Animation Ext: Select name.#.ext. This specifies that the filenames will have the format *prefix.frameNumber.fileFormat.* For example, batch rendering the entire 240-frame animation will create fish.0001.iff, fish.0002.iff, and so on through fish0240.iff.
   c. Start Frame: Enter 0, the first frame of the animation sequence to be batch rendered.

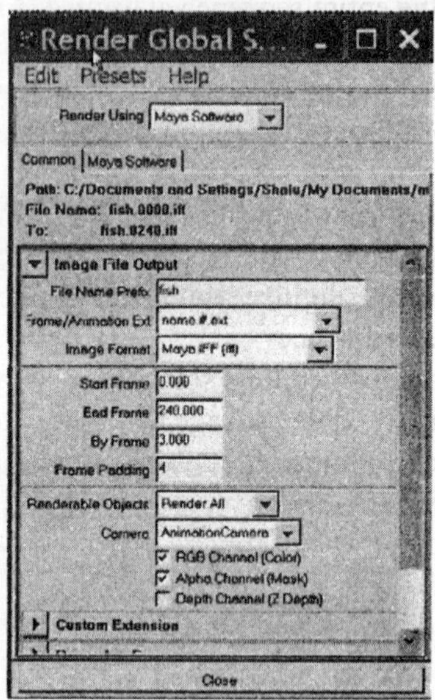

d. End Frame: Enter 250, the last frame to be batch rendered.

e. Frame Padding: Enter 4. This causes the *frameNumber* part of the filenames to be four digits prefixed with 0s. For example, the filenames will fish.0001.iff tinstead of fish.1.iff.

f. Set the renderable Camera to be *AnimationCamera*. This is the camera which we will be rendering.

3. For the remaining options in the Render Globals, you'll use the settings you defined from Chapter 8. Maya will render using resolution (640x480), and anti-aliasing quality (Production Quality) with ray tracing on.

From the menu bar, choose *Render>Batch Render* to kick off your render job. It will take a while-maybe even hours-so you take a tea break before you view the final rendered animation.

11.3 Post-production

The post production phase is when editorial comes in. Editors pick up the animated frames and compile them together to make the final movie. The sound track, complete with narration, music, and sound effects, is added into the clip. If the animation is to be mixed in with live action footage then the live action footage is digitized and composited with the animated sequence as well.

The key to all of this is non-linear computer editing.

In traditional video editing folks would copy segments of your tape to a master tape, laying shots in order linearly onto the tape. If you got to the end of a sequence and decided that you wanted to change the third shot, you would have to edit the entire sequence all over again. Grrr!

Non-linear computer editing allows you manipulate movie clips and audio (animated or otherwise) in the same way that you would add and change words in a word-processor: simply Cut, Copy and Paste the shots to stitch your movie together.

If you want to do this at home, you will need to get a video capture card, which will allow you to sample and store clips of your video to hard disk. You will need to digitize your clips frame by frame to composite them with an animated sequence.

Non-linear editing software usually comes bundled with the card. Get a powerful, flexible package that is easy to work with. Some worthwhile packages are Adobe Premiere, Ulead MediaStudio, Judgement, EditDV and Avid. .The software will allow you to edit the audio tracks adding voice overs and background music. You can then use it to the tape the final clip onto film or video as desired.

For example, in our fish animation, we could use a live-action footage of people drinking at a party. The footage would need to be digitized into a sequence of frames. These frames could then be used as images to be reflected in our mirror (achieved by compositing the live action image and the animated image with an appropriate mask). This would give the effect of a room filled with live people. Care must be taken that the lighting of each of the two scenes blends with the other to make the composite look believable.

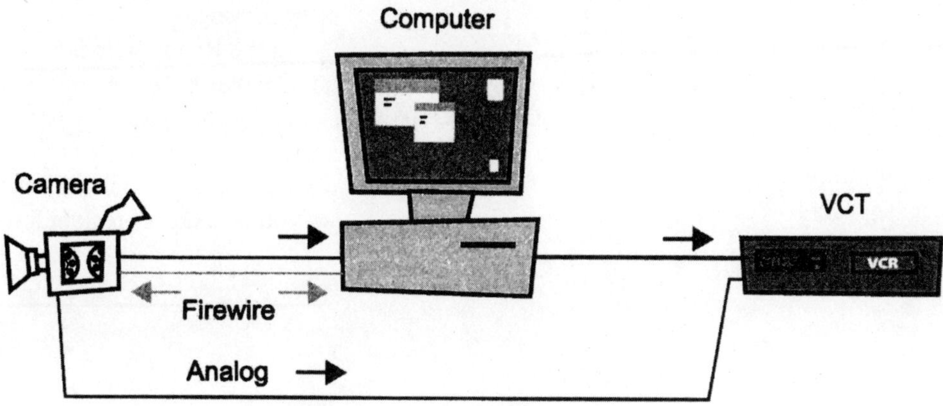

Fig.11.7: Non-linear editing

Finally, using the non-linear software, you can add in sounds effects and the final movie clip can be produced and transferred directly to VHS.

The steps followed by professional editorial departments are analogous to the ones described above. The exact mechanics of non-linear editing is beyond the scope of this book-maybe in Part 2!

The final images are stored in the directory specified by the Path variable we noted earlier. Go to this directory. You will see the individual images of the

11.4 Finally: Our Movie

animation saved as fish.*.iff. Once Maya is done generating all the images of the animation, double click on the fish.0000 image. This will bring up the Fcheck program that Maya uses to display images and will display the 0th frame of our render. Choose *File>Open Animation* and open fish.000.iff again. This will display the rendered animation of your film. Enjoy!

ColorPlate shows some sample images from our movie. This animated world is provided for you as well under *Models/worldwithfish.mb.*

Summary

In this chapter, we have developed a movie following the same processes that the professionals use. We went from pre-production, where we defined out fish story; to production, where we animated the movie; and finally to post-production. We animated and finally rendered a stunning movie using Maya.

This chapter ends our journey into 3D Graphics. We hope you have enjoyed your experience and are motivated to create your own 3D graphics, using the techniques you have learned in this book.

Appendix A

OpenGL and GLUT

What is OpenGL?

OpenGL seems to be as close to an industry standard as there is for creating 2D and 3D graphics applications. OpenGL provides a library of graphics functions for you to use within your programming environment. It provides all the necessary communication between your software and the graphics hardware on your system. The OpenGL API is a portable API that can be compiled and run on many platforms.

OpenGL programs are typically written in C and C++.

What is GLUT?

GLUT (pronounced like the glut in gluttony) is the OpenGL Utility Toolkit, a window system independent toolkit for writing OpenGL programs. It implements a simple windowing application programming interface (API) for OpenGL. GLUT makes it considerably easier to learn about and explore OpenGL programming. GLUT is designed for constructing small to medium sized OpenGL programs. While GLUT is well-suited to learning OpenGL and developing simple OpenGL applications, GLUT is not a full-featured toolkit so large applications requiring sophisticated user interfaces are better off using native window system toolkits. GLUT is simple, easy, and small. The GLUT source code distribution is portable to nearly all OpenGL implementations and platforms. GLUT is not open source. Mark Kilgard maintains the the copyright.

What libraries and header files will I need to compile and link OpenGL/GLUT programs?

To compile and link OpenGL programs, you'll need the OpenGL header files and libraries.

• Windows: If you are running Windows 98/NT/2000 then this library has already been installed in your system. Otherwise download the Windows

OpenGL library from Microsoft at:
ftp://ftp.microsoft.com/softlib/mslfiles/opengl95.exe
• Mac OS X: If you are running MacOS X then download the OpenGL SDK from Apple at:
http://developer.apple.com/opengl/
• LINUX/BSD: If you are running LINUX then download Mesa - the OpenGL work-alike at:
http://www.mesa3d.org/
• UNIX: The OpenGL libraries are available directly from your UNIX vendor

The OpenGL Utility Toolkit (GLUT) is available for download at:
http://www.opengl.org/resources/libraries/glut.html
Note that you only need to download the glutdlls.zip archive, but you may also want to download the source code and the HTML version of the API.

To run OpenGL programs that take advantage of OpenGL 1.2 or above, you will need to download vendor-specific OpenGL Drivers for your particular graphics card. Refer to:
http://www.opengl.org/documentation/spec.html/
for more details.

Where do I install these files?

Microsoft Windows 95 and above:
If you're using Visual C++ under Windows 9x, NT or 2K, your compiler comes with include files for OpenGL and GLU, as well as .lib files to link with. Install glut.h in your compiler's include directory, glut32.lib in your compiler's lib directory, and glut32.dll in your Windows system directory (c:\windows\system for Windows 9x, or c:\winnt\system32 for Windows NT/2000). In summary, a fully installed Windows OpenGL development environment will look like this:

File	Location
gl.h glut.h glu.h	[compiler]\include\gl
Opengl32.lib glut32.lib glu32.lib	[compiler]\lib
Opengl32.dll glut32.dll glu32.dll	[system]

where [compiler] is your compiler directory (such as c:\Program Files\Microsoft Visual Studio\VC98) and [system] is your Windows 9x/NT/2000 system directory (such as c:\winnt\system32 or c:\windows\system).

You'll need to instruct your compiler to link with the OpenGL, GLU, and GLUT libraries. In Visual C++ 6.0, you can accomplish this with the Project menu's Settings dialog box. Scroll to the Link tab. In the Object/library modules edit box, add glut32.lib, glu32.lib, and opengl32.lib to the end of any text that is present.

For UNIX or UNIX-like operating systems:
If you don't find the header files and libraries that you need to use in standard locations, you need to point the compiler and linker to their location with the appropriate -I and -L options. The libraries you link with must be specified at link time with the -l option; -lglut -lGLU -lGL -lXmu -lX11 -lm is typical.

For Mac OS X:
Apple provides excellent documentation on how to access header and library files and code using their OpenGL SDK at:
http://developer.apple.com/opengl/

For Linux and other systems:
You can find excellent documentation for this at:
http://www.mesa3d.org/

More OpenGL Documentation

If you would like some more information, then I would suggest you spend time browsing the OpenGL main web site at:
http://www.opengl.org:
You will find all related documentation and supporting libraries on this web-site.

Appendix **B**

Downloading and running Sample Code

How do you download and install the sample code?

• Download the sample code (in zipped format) from:
http://www.springeronline.com/0-387-95504-6

Most of our code is portable to UNIX or any platform with OpenGL installed.

• Go to the directory where you want to install the sample code.
• Unzip the downloaded zip file into this folder.

You should now see the following folders and files installed:

• *OpenGL/*: contains the examples referred to in the book, grouped by the example number that they are used within. For e.g., under the folder: *OpenGL/Example1_1/* you will find the source code for *Example1_1* in the file *Example1_1.cpp*.

• *OpenGL/*: contains the code for the utility functions:

- Vector utilities: *vector.h*
- BMP image reading and writing: *bmp.cpp*, *bmp.h*
- VRML models reading: *vrml.h*, *vrml.cpp*
- Interpolation routines: *linearinterpolation.cpp*, *linearinerpolation.h* *cubicinterpolation.cpp*, *cubicinterpolation.h*

• *OpenGL/Models/*: contains the VRML and Alias models used in our examples.

• *OpenGL/Images/*: contains the images used as textures in our examples.

How do you compile and link the programs?

We detail this process for Microsoft Visual C++ compiler. Any other visual compiler will be similar. UNIX/Linux users will need to compile and link their programs from the command line or from a MakeFile as appropriate (See Appendix A).

1. Startup Microsoft Visual Studio
2. Create a Project

To create an Open GL project, we are going to need to modify one of the standard projects defined by Microsoft's Visual Studios.

Select *File* > *New* from the main menu. Next select the *Projects* tab and select the *Win32 Console Application* project. Don't forget to enter the project name and the path to the project folder.

3. Specify The Settings

Select the *Project>Settings* from the Main Menu.
Select the *Link* tab. This allows you to edit some of the linking options of the compiler.
Select the *General* category from the drop down list.
Select the *All Configurations* option from the list. This will make our settings valid for both compiling the project in Debug mode as well as compiling the project in Build mode.
Finally, add the glut32.lib, glu32.lib and opengl32.lib to the *Object/library modules* box.
Click the *OK* button to save the changes.

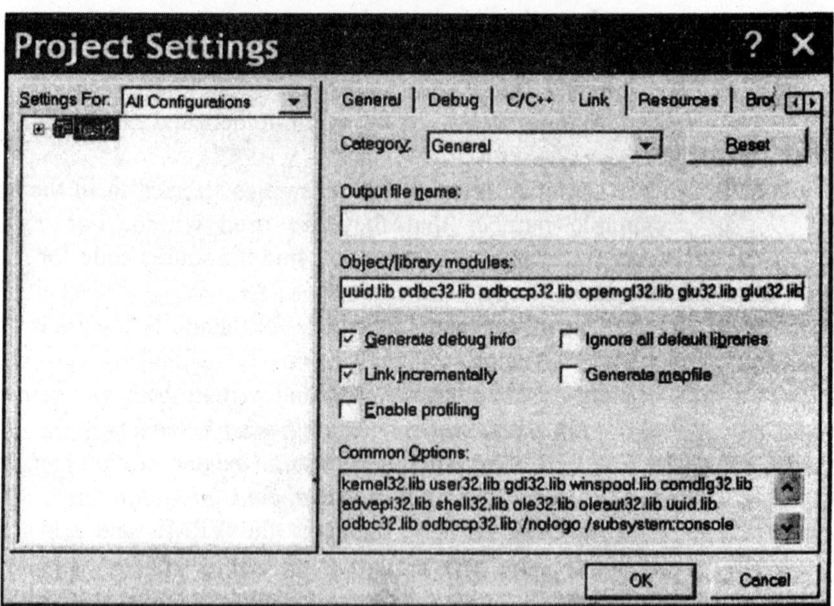

4. Add the necessary files to the Project

The next step is to add the required .cpp (or .c) files to our project so we can test the installation. Select *Project > Add to Project>Files* from the Main Menu. For this example, select our file *Example1_4/Example1_4.cpp*. Visual Studios will automatically add this to your project and save it in your project's folder. Follow this step to add more files if needed.

5. Compile and Execute
Finally, press Ctrl-F5 to build and execute the program. *Example1_4* should look something like this:

Enjoy!

Appendix C

Maya Personal Learning Edition (PLE)

What is Maya?

Academy Award winning Maya software is the world's most powerfully integrated 3D modeling, animation, effects, and rendering solution. Maya also adds to the quality and realism of 2D graphics. That's why film and video artists, game developers, visualization professionals, Web and print designers turn to Maya to take their work to the next level.

What is Maya PLE?

Maya Personal Learning Edition is a special version of Maya software, which provides free access to Maya for non-commercial use. It gives 3D graphics and animation students, industry professionals, and those interested in breaking into the world of computer graphics (CG) an opportunity to explore all aspects of the award winning Maya software in a non-commercial capacity.

Where do I download the software?

You can download the software from the following URL
http://www.alias.com/eng/products-services/maya/maya_ple/

Follow the instructions on how to install the program onto your desktop.
This page also provides more details on software and hardware requirements for running the program.

Bibliography

What follows is a bibliography in computer graphics. In addition to being a list of references from the various chapters, it also contains references to books and journals where the latest research in the field is being published.

ADAM91 **Adams, L.** *Supercharged C++ Graphics*, WindCrest, 1991

APOD99 **Apodaca,A.A., L. Gritz.**, *Advanced RenderMan: Creating CGI for Motion Pictures*, Morgan Kaufmann,1999

BADL91 **Badler, N.I., B.A. Barksy, D. Zeltzer (ed.),** *Making Them Move: Mechanics, Control, and Animation of Articulated Figures,* Morgan Kaufman, 1991.

BART87 **Bertels, R., J. Beatty, B. Barsky**, *An Introduction to Splines for Use in Computer Graphics and Geometric Modeling*, Morgan Kaufmann, 1987.

BOVI00 **Bovik, A.**, *Handbook of Image and Video Processing*, Academic Press, 2000

COHE01 **Cohen, E., R. F. Riesenfeld, G Elber**, *Geometric Modeling with Splines: An Introduction*, AK Peters, Ltd., 2001.

CUNN92 **Cunnighamn, S., N. Knolle, C. Hill, M. Fong, J. Brown**, *Computer Graphics using Object Oriented Programming*, John Wiley and Sons, 1992.

DEBO01 **deBoor C.**, *A Practical Guide to Splines*, Springer-Verlag, 2001

DERA03 **Derakhshani, D.,** *Introducing Maya 5: 3D for Beginners*, Sybex Inc, 2003

DERO98 **DeRose, T., M. Kass, T. Truong**, *Subdivision Surfaces in Character Animation*, ACM Proceedings of the 25th annual conference on Computer graphics and interactive techniques, pp.: 85-94

DUTR03 **Dutre, P., P. Bekaert, K. Bala**, *Advanced Global Illumination*, AK Peters, Ltd., 2003

FARR97 **Farrell, J.A,** *From Pixels to Animation*, Academic Press, 1997

FOLE82 **Foley, J., A. van Dam**, *Fundamentals of Interactive Computer Graphics*, Addison-Wesley, 1982

FOLE93 **Foley, J.D., A. van Dam, S.K. Feiner**, *Introduction to Computer Graphics*, Addison-Wesley Professional; 1993

FOLE95 **Foley, J.D., A. van Dam, S.K. Feiner, J.F. Hughes**, *Computer Graphics: Principles and Practice in C*, Addison-Wesley Professional; 1995

GLAE99 **Glaeser, G.H. Stachel**, *Open Geometry: OpenGL + Advanced Geometry*, Springer Verlag, 1999

GLAS89 **Glassner, A.S.**, *An Introduction to Ray Tracing*, Morgan Kaufmann, 1989.

GLAS04 **Glassner, A.S.**, *Interactive Storytelling: Techniques for 21st Century Fiction*, AK Peters, Ltd., 2004

GLAS95 **Glass, G., B. L. Schuchert**, *The STL Primer*, Prentice Hall PTR; 1995

GOUR71 **Gouraud, H.**, *Continuous Shading of Curved Surfaces*, IEEE Transactions on Computers, C- 20(6), June 1971, pp. 623-629.

HILL00 **Hill, F.S.**, *Computer Graphics Using OpenGL*, Prentice Hall, 2000

LAMO03 **LaMothe, A.**, *Tricks of the 3D Game Programming Gurus-Advanced 3D Graphics and Rasterization*, Sams; 2003

LAVE03 **Lavender, D.**, *Maya Manual*, Springer Professional Computing Series, 2003

LASS87 **Lasseter, J.**, *Principles of Traditional Animation Applied to 3D Computer Animation*, SIGGRAPH 87, pp. 35-44

LENG93 **Lengyel, E.**, *Mathematics for 3D Game Programming and Computer Graphics*, Charles River Media; 2003

LISC03 **Lischner, R.**, *STL Pocket Reference*, O'Reilly & Associates, 2003

MAGN85 **Magnenat-Thalmann N., D. Thalmann**, *Computer Animation. Theory and Practice*, Springer Verlag, Tokyo, 1985

NEWM81 **Newman W.M., R.F. Sproull**, *Principles of Interactiye Computer Graphics*, McGraw-Hill, 1981

PAIS98 **Pai, S.G., R. Pai**, *Learning Computer Graphics*, Springer Verlag, 1998

PARE01 **Parent, R.**, *Computer Animation: Algorithms and Techniques*, Morgan Kaufmann, 2001

PORT84 **Porter T., and T. Duff**, *Compositing Digital Images*, SIGGRAPH 84, pp. 253-259.

PROS94 **Prosise, J.**, *How Computer Graphics Work*, Ziff-Davis Press, Emeryville, CA, 1994.

REEV83 **Reeves, W.T.**, *Particle Systems-A Technique for Modeling a Class of Fuzzy Objects*, SIGGRAPH 93, pp. 359-376.

ROST04 **Rost, R.J., J. M. Kessenich, B. Lichtenbelt, M. Olano**, *Open Gl Shading Language*, Addison-Wesley Professional, 2004

SHIR02 **Shirley, P**., *Fundamentals of Computer Graphics*, AK Peters, Ltd., 2002

SHRE03 **Shreiner,D., M. Woo, J. Neider, Tom Davis**, *OpenGL Architecture Review Board, OpenGL Programming Guide: The Official Guide to Learning OpenGL, Version 1.4*, Addison-Wesley Professional, 2003

THOM95 **Thomas, F., O. Johnston**, *Disney Animation, the Illusion of Live*, Disney Editions; revised edition, 1995

WARR01 **Warren, J., H. Weimer**, *Subdivision Methods for Geometric Design: A Constructive Approach*, Morgan Kaufmann, 2001

WATT93 **Watt A.**, *3D Computer Graphics*, Addison Wesley, 1993.

Index of Terms